Process Development Division 2017

Core Programming Area at the 2017 AIChE Annual Meeting

Minneapolis, Minnesota, USA
29 October - 3 November 2017

ISBN: 978-1-5108-5800-8

Printed from e-media with permission by:

Curran Associates, Inc.
57 Morehouse Lane
Red Hook, NY 12571

Printed by Curran Associates, Inc. (2018)

For permission requests, please contact AIChE
at the address below.

AIChE
120 Wall Street, FL 23
New York, NY 10005-4020

Phone: (800) 242-4363
Fax: (203) 775-5177

www.aiche.org

Additional copies of this publication are available from:

Curran Associates, Inc.
57 Morehouse Lane
Red Hook, NY 12571 USA
Phone: 845-758-0400
Fax: 845-758-2633
Email: curran@proceedings.com
Web: www.proceedings.com

TABLE OF CONTENTS

Some abstracts in this book refer to images that were not made available by the publisher

496260 Addressing the Downstream Challenges of Viral Therapy for Cancer Treatment – a Study of Rhabdoviral Vector Purification

Abstract:

Oncolytic viruses (OVs) are a class of cancer therapies that are currently going through clinical trials on their way to full approval. For example, the world's first clinical trial using a combination of two viruses to kill cancer cells and stimulate an immune response is currently being conducted. Existing downstream purification processes use a combination of chromatography and membrane-based processes to remove both process-related (e.g. cell debris, host cell proteins, and nucleic acids) and product-related impurities (e.g. aggregated and empty virus particles). One of the main challenges during viral vector manufacturing is the mandatory final sterile filtration step. The primary issue is the severe membrane fouling, as indicated by a dramatic and uncontrollable transmembrane pressure increase associated with the low recovery of the desired final product - recent studies reporting up to 80% losses in the total vector titer. We have evaluated the fouling behaviour of commercially available sterile filters with various membrane morphologies using a Rhabdoviral vector.

As a potential improvement in the downstream process, we have investigated the use of monolithic column chromatography. This technology features a continuous stationary phase with a network of interconnected pores that has been previously shown to be ideal for the production of 'large' biomolecules. The monolithic column (1 mL capacity) made of polymethacrylate copolymers functionalized with quaternary ammonium groups (BIA Separations) was evaluated using NGC Quest Plus chromatography system (BioRad). The flow rate and buffer composition conditions used for the monolith chromatography were optimized in terms of the recovery of the Rhabdoviral vector. At a relatively high flow rate of 4 ml/min, the purified Rhabdoviral vector sample was purified using a step-elution salt concentration gradient with the majority of impurities washed out at lower salt concentrations. To investigate the fouling propensity of the different biomolecules that were collected in each chromatographic fraction, a series of small-scale (13mm polycarbonate filter holder with the filter area of 0.5 cm^2) sterile filtration studies were conducted. The resulting TMP profiles for each fraction (at constant flux) were analyzed using various pore plugging/blocking models. Our current hypothesis is that by increasing removal of the process- and product-related impurities using the monolithic column, we can improve the sterile filtration process performance in terms of throughput and product recovery that is needed for the large-scale production of oncolytic viruses.

497681 A Small Molecule Based Affinity Membrane Spin Column for Antibody Purification

Abstract:

Antibodies have extraordinary specificity and affinity to antigens, which in turn makes them important candidates to be used in numerous applications including clinical, academic and industrial, while new implementations are continuously being explored in academic research. A major contributor to the cost of antibody applications is mostly linked to the downstream production process, or more specifically the usage of Protein A (or G) affinity columns for purification of antibodies. These columns are expensive and have short life cycles with several obstacles that preventing them to be used repeatedly. Here, we design a novel affinity spin column for the purification of monoclonal and polyclonal antibodies from complex protein mixtures such as cell culture media of hybridomas and ascites fluids. This method utilizes the nucleotide-binding site (NBS) that is located on the Fab variable domain of immunoglobulins to enable capturing of antibody molecules on a membrane affinity column via a small molecule, which has a moderate binding affinity to the NBS. A ring structured small molecule was attached to the regenerated cellulose membrane in order to target the NBS for capturing antibodies on the spin column. Antibody capture was accomplished by running of samples in an equilibration buffer (20 mM TRIS pH 7.0) through the spin column at 500 rpm and elution of the antibody was achieved by running a mild elution buffer (0.5 M KCl in 20 mM TRIS pH 7.0). Several pharmaceutical antibodies were tested on the spin column, and the results indicate that the column efficiency for antibodies was >95%, with a purity level of >95%. Furthermore, small molecule based affinity column has a strong binding specificity towards antibodies, and has a dynamic binding capacity of ~48 μg/mg which is greater than any current protein A (or G) spin columns. The quality and the capacity of this small molecule affinity membrane spin column method is further evaluated for a number of parameters such as: injection concentrations, volumes, number of membranes, antibody/protein-contaminant combinations, and column reusability and stability. This study provides a good tool for fast antibody purification at a small scale.

486714 Removal of Cyanobacterial Toxins Using Polymeric Nanoparticles

Abstract:

Communities are faced with significant threats to health by the supplied water from lakes or reservoirs that develop algal bloom. During a bloom, cyanobacteria release the bacterial cell wall component endotoxin, and other toxins into the water, which enter the bloodstream upon ingestion causing an overwhelming inflammatory response in humans. Water treatment plants currently use a series of standard cleansing procedures to diminish cyanotoxins, however, finding an effective and affordable technique remains a challenge. Here, we propose a new approach based on biocompatible polymeric NPs in particulate, and membrane forms with a previously proven capability to achieve improved effectiveness and efficiency in removing cyanotoxins from water. Removal of cyanotoxins were assessed using nanometer-sized poly-e-caprolactone (PCL) NPs with specifically different diameters and shapes as synthesized in the PIâ€™s laboratory. Membranes were constructed incorporating PCL NPs of various diameters and shapes to create a corrugated surface for effective cyanotoxin removal from contaminated water.

486597 Increased Production of Yersiniabactin (Ybt) and the Anthranilate Analog through Media Optimization and Ybt Metal Binding Characterization

Abstract:

Yersiniabactin (Ybt) is a mixed nonribosomal peptide-polyketide natural product that binds a wide range of metals with the potential to impact processes requiring metal retrieval and removal. The heterologous production of Ybt and an associated anthranilate analog are increased through systematic screening and optimization of culture medium components. Specifically, a Plackett-Burman design-of-experiments methodology was used to screen 22 components and to determine those contributing most to siderophore production. L-cysteine, L-serine, glucose, and casamino acids significantly contributed to the production of both compounds. Ybt and the anthranilate analog titers were increased to 867 ± 121 mg/L and 16.6 ± 0.3 mg/L, respectively, an increase of ~3,800% and ~7,900% relative to production in M9 medium.

The range of metals that Ybt forms a complex with were determined by using mass spectrometry. Using LC-MS the analysis of water samples exposed to aqueous and surface-localized Ybt, quantitative assessment of binding was completed with metals that included palladium, magnesium, and zinc. In total, Ybt showed affinity for 10 metals. Next, Ybt-modified XAD-16N resin (Ybt-XAD) was utilized to quantify the degree of metal binding and metal removal abilities, showing a rank order of Fe>Ga>Ni>Cu>Crâ‰ˆZn>Co>Pd>Mg>Al. Batch tests were conducted using Ybt-XAD for the treatment of wastewater from a local precious metal plating company, showing selective removal of nickel from the aqueous effluent.

499390 Applications of Magnetically Responsive Micro/Mesoporous Carbon Adsorbents in Biorefining

Abstract:

Inhibitory compounds commonly arise at multiple points in the biorefining pipeline, including with respect to both toxic byproducts in the feedstock as well as the final target product itself. In both cases, if not effectively removed, their presence can have deleterious effects on the microbial biocatalyst and, in turn, its overall performance. Here we demonstrate how a series of novel, magnetically responsive micro/mesoporous carbon powder (MMCP) adsorbents can be employed for the facile and efficient removal of inhibitory bioproducts and byproducts alike. In addition to large surface areas (>1400 m^2/g) promoting high adsorption capacities, said MMCPs also uniquely include magnetic Ni nanoparticles that enable the direct and facile magnetic retrieval of spent adsorbents from the culture broth, bypassing the need for column configurations (e.g., packed or expanded bed). As a preliminary demonstration of concept, the use of MMCPs for *in situ* removal of n-butanol has been studied under model conditions. Although it represents a promising biofuel, n-butanol production by conventional batch fermentation is limited as a result of its cytotoxic effects. Butanol loading capacities of a family of MMCPs (with 4-10%wt. Ni content) were found to not be significantly impacted by Ni content and, furthermore, performed comparably to other, previously-studied commercial resins and activated carbons. Magnetic retrieval of the mesoporous powder is dependent on the Ni content, with up to 89% retrieval achieved in 6 min with 10%wt. Ni. Desorption studies using retrieved adsorbents demonstrated an average of 93% recovery of the total adsorbed butanol. Kinetic studies, meanwhile, indicate that neither the effective adsorption or desorption rates should constitute a bottleneck with respect to the future development of semi-continuous butanol fermentation processes using MMCPs. Finally, the further utility of MMCPs for the upstream removal of toxic byproducts derived from the pretreatment of lignocellulose has been investigated. Furfural, for example, can be readily removed using MMCPs that are subsequently retrieved from the culture. Said pretreatment steps improve the fermentability of biomass-derived sugar mixtures, supporting improved production of ethanol and lactate by engineered *Escherichia coli*.

501387 Improving Enzymatic Synthesis of β-Lactam Antibiotics By in-Situ Crystallization

Abstract:

Improving enzymatic synthesis of β-lactam antibiotics by in situ crystallization

Matthew A. McDonald*, Lukas Bromig*, Andreas S. Bommarius*, Martha A. Grover*, and Ronald W. Rousseau*

**School of Chemical & Biomolecular Engineering, Georgia Institute of Technology, Atlanta, Georgia*

Keywords: reactive crystallization, β-lactam antibiotics, penicillin G acylase

In 2010 the two most common classes of β-lactam antibiotics, the penicillins and cephalosporins, accounted for nearly 60% of world antibiotic consumption, with delivery of over 20 billion doses of penicillin-derived antibiotics.[1] Traditionally, β-lactam antibiotics have been made batch-wise in a process requiring several chemically intensive steps, such as amine protection by benzyl chloroformate and deprotection by hydrogenation with palladium catalysts, resulting in large amounts of potentially contaminated waste.[2] The contaminated waste poses two threats: the spread of antibiotic resistance in wastewater treatment plants[3] and inadvertent exposure of sensitive individuals to a sever allergen. A 2012 study found that 92 out of 11,761 patients at Mount Sinai Hospital in New York City suffered anaphylaxis (a rapid, potentially deadly allergic reaction) from penicillin, a prevalence of 0.7%.[4]Enzymatic synthesis of β-lactam antibiotics can help to mitigate these risks by operating in mild aqueous conditions. The enzyme used industrially for this purpose is penicillin G acylase (PGA); it also catalyzes the hydrolytic degradation of the antibiotic (termed secondary hydrolysis), making the desired product an intermediate. In-situ product isolation can prevent the enzyme from degrading the antibiotic.

We use ampicillin and cephalexin as a model penicillin and cephalosporin, respectively. PGA catalyzes the condensation of 6-aminopenicillanic acid (6-APA) and phenylglycine methyl ester (PGME) into ampicillin and 7-aminodesacetoxy-cephalosporanic acid (7-ADCA) and PGME into cephalexin. Ampicillin and cephalexin have lower saturation concentrations than 6-APA and 7-ADCA, respectively, making it possible to supersaturate the solution with respect to the antibiotic. The supersaturation can then be used to crystallize pure ampicillin and cephalexin, isolating it from PGA.[5]Degradation of ampicillin and cephalexin by PGA yields phenylglycine, a sparingly soluble contaminant. PGA also hydrolyzes the highly soluble PGME into phenylglycine (termed primary hydrolysis). To maintain a pure solid phase phenylglycine must not be allowed to reach its saturation concentration. Typical crystallization processes utilize evaporation, cooling, and other chemical process to generate supersaturation. In this process, PGA generates the supersaturation necessary for the separating phase change (from solute to crystal).

To better understand the mechanism and quantitatively predict the process behavior, a model of the enzyme kinetics, including the synthesis, primary, and secondary hydrolysis reactions, was developed. Existing kinetic models for ampicillin and cephalexin synthesis by PGA do not consider the effects of pH value on the reaction despite the fact that pH is known to affect the enzyme selectivity and activity. The pH value is shown to decrease with increasing conversion because of the generation of an acid group in phenylglycine. We have improved on previous models such that our model considers the effect of pH and removal of species from solution via crystallization.[6]

For ampicillin the new process showed maximum improvements in yield and selectivity of 33% and 49% over the non-crystallizing control, respectively, all while maintaining a strict purity constraint. However, these two maxima cannot be simultaneously achieved; instead selectivity is optimized at a lower pH value around 6 where the enzyme is suspected to bind 6-APA more strongly, and yield improves at a higher pH value around 7 where catalytic activity is improved but less specific.[6]Preliminary results show cephalexin enjoys similar benefits from reactive crystallization.

More recently we have explored continuous reactive crystallization to synthesize β-lactams because it may enable greater efficiency and access to different chemistries. To operate a continuous process the stability and longevity of the enzyme must be characterized. We have determined the temperature activity profile and the empirical parameter T^{60}_{50}, the temperature at which 50% of the initial activity degrades within one hour, of an industrially relevant PGA, Assemblase® from DSM-Sinochem. . Temperature ramp studies were also used to determine catalyst longevity (in the form of total turnover number, TTN). Continuous operation could increase conversion over batch by tailoring the conditions to optimize reactive crystallization; for example, a plug flow reactor/crystallizer can operate with excess 6-APA or 7-ADCA relative to PGME to hinder hydrolysis of the PGME while a recycle stream could act to seed the plug flow apparatus to slow antibiotic hydrolysis via crystallization. The aforementioned model is combined with enzyme stability data to map the large design space opened by continuous operation.

References

[1] Gelbrand, H., Miller-Petrie, M., Pant, S., Gandra, S., Levinson, J., Barter, D., White, A., Laxminarayan, R., Ganguly, N. K., and Kariuki, S. (2015) The State of the World's Antibiotics 2015, *Wound Healing Southern Africa 8*, 30-34.

[2] Giordano, R. C., Ribeiro, M. P., and Giordano, R. L. (2006) Kinetics of β-lactam antibiotics synthesis by penicillin G acylase (PGA) from the viewpoint of the industrial enzymatic reactor optimization, *Biotechnology advances 24*, 27-41.

[3] Deschamps, E., Vasconcelos, O., Lange, L., Donnici, C. L., Silva, M. C. d., and Sales, J. A. (2012) Management of effluents and waste from pharmaceutical industry in Minas Gerais, Brazil, *Brazilian Journal of Pharmaceutical Sciences 48*, 727-736.

[4] Albin, S., and Agarwal, S. Prevalence and characteristics of reported penicillin allergy in an urban outpatient adult population, 6 ed., pp 489-494, OceanSide Publications, Inc.

[5] Encarnación-Gómez, L. G., Bommarius, A. S., and Rousseau, R. W. (2016) Reactive crystallization of β-lactam antibiotics: strategies to enhance productivity and purity of ampicillin, *Reaction Chemistry & Engineering 1*, 321-329.

[6] McDonald, M. A., Bommarius, A. S., and Rousseau, R. W. (2017) Enzymatic reactive crystallization for improving ampicillin synthesis, *Chemical Engineering Science 165*, 81-88.

490473 Separation of Fatty Acids and Fatty Acid Esters Using Nanoporous Polymeric Membranes

Abstract:

Membrane separations are typically inexpensive and can be scaled to industrial quantities, but the development of membranes that can separate organic chemicals lag behind those that separate gasses and water. Part of the challenge in this field is that the polymeric membranes swell in the presence of organic chemicals and the swelling impacts their abilities to separate chemicals. Despite these challenges, the potential applications of membranes that separate organic liquids and solids is impressive. For instance, worldwide over 200M tons of vegetable, tall, and fish oils are isolated each year, and their fatty acids are biorenewable, safe feedstocks for the chemical community. Unfortunately, the purification of fatty acids using traditional methods such as distillation, selective precipitation, and chromatography are energy intensive or not readily scaled to separate unsaturated fatty acids. Commercial applications of fatty acids requires starting with a mixture of fatty acids that can vary based on which season the crops were grown and the plant source for the oils.

We recently developed a series of membranes based on polyepoxies that can separate some of the key fatty acids found in vegetable and fish oils. These membranes were synthesized by the addition of diamines and di or triepoxides that react to yield highly cross-linked membranes. The selectivity of the membranes can be tuned by the selection of the epoxide and diamines used in their synthesis. These membranes were able to separate saturated and unsaturated fatty acids.

509411 Predicting Temperature-mediated Solid Form Transformations in Small Molecule Crystals with Molecular Dynamics

Abstract:

Many pharmaceutical compounds can crystallize into more than one solid form, and the overall solubility and bioavailability of a therapeutic compound can vary markedly between different structures. Selecting a stable and soluble solid form is therefore an important step in the drug development process, and an unintended late-stage transformation into a new solid form can severely delay a therapeutic reaching the market. Computational models of a compoundâ€™s solid form landscape can reveal previously unobserved crystal structures and assist in selecting the appropriate form for commercial manufacturing. However, the current approaches for crystal structure predictions (CSP) often neglect the effects of temperature on solid form stability. Such approaches cannot capture temperature-mediated transformations that can occur during product manufacturing and storage.

In this work, we present a method to compute the temperature-dependent stability of small molecule crystals and capture temperature-mediated solid form transformations. All crystal structures of interest are heated from zero Kelvin up to ambient conditions using molecular dynamics simulations. The sets of simulations are combined to construct temperature-dependent free energy profiles for each structure. We demonstrate this technique on multiple polymorphic small molecules of pharmaceutical interest, and show that current point-charge potentials are capable of capturing the large entropy differences between forms which facilitates crystal transformations at high temperature.

In addition, we show that molecular dynamics simulations can yield visual insights into when crystalline systems undergo critical property changes as a result of a change in temperature. These changes include solid-solid transformations, order-disorder transitions, and lattice minima interconversions.

509412 Polymer D-255 Cement Fluid Loss Additive Control Optimization

Abstract:

Incorporated and optimized new designs with D-255 FLAC technology to meet the same performance given by previous additive D-167. The project had the main goal to improve control of water loss from cement to formation by reducing the permeability of the cement filter cake, and increasing viscosity of the aqueous phase. The testing results concluded new technology to have better stability and compressive strength development. The replacement of new technology with a PC/PR of 0.27 will save $81K to Schlumberger with a 50% sales profit increase in five years.

Modeling Paste Transport Systems for FGD Water Disposal

Trent Rogers

Research & Development

SOUTHERN COMPANY

What is Paste?

- Fly ash encapsulated FGD wastewater
- Non-Newtonian, cementitious fluid pumped into landfills
- FGD wastewater disposal technique that is zero liquid discharge (ZLD) compliant

Uncured paste sample vs cured paste sample

Project Overview

- Model the approximate cost of operating a paste system using a rheometer
- Two dominant cost factors:
  1. Energy/capital costs of pumping (head loss)
  2. Raw material cost of paste (mix design)
- No reliable model for rheopectic pseudo-plastics in mining industry

Mining paste backflow system

Rheometer Testing

- Used to characterize non-Newtonian flow
- Two rheometer measurements
  1. Shear rate: angular velocity of spindle
  2. Shear stress: torque of motor
- These measurements can be used to simulate pipe wall interaction

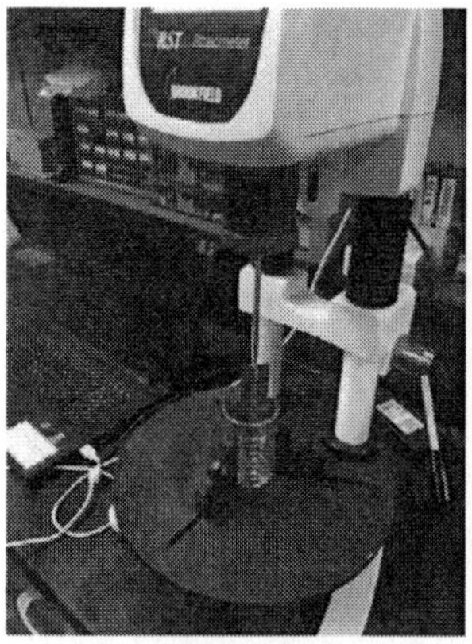

Rheometer Setup

Estimating Head Loss

- Modeling head loss using a rheometer is fairly common; time dependency is often neglected
- Paste is an "apparent rheopectic" fluid due to rapid cementitious reaction
- Modified cement slurry model (Haimoni and Hannent) for rheopectic behavior

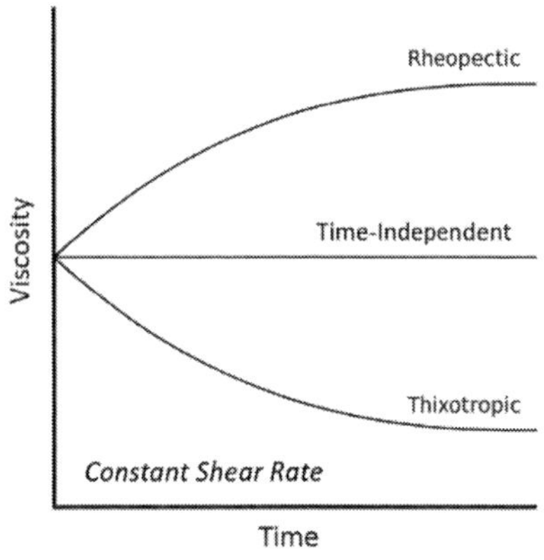

Mix Design Costs

- Chemical additives greatly affect rheology
- Balancing material cost and energy cost is essential
- Mix design can also impact capital costs if a different pump is required

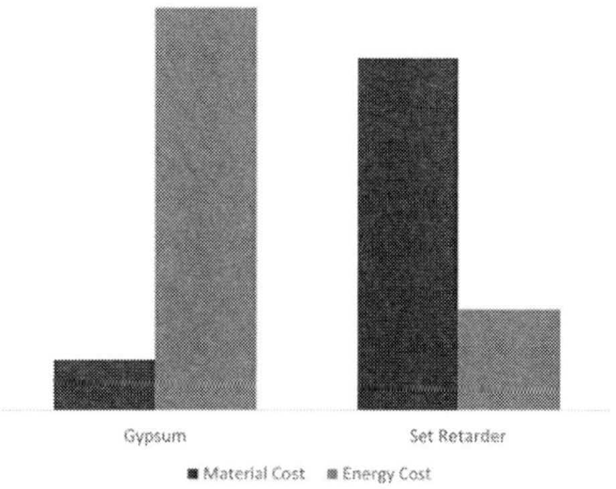

Summary of Modeling Process

1. *Rheometer*
   - Use a designed velocity and rheometer geometry to determine a controlled shear rate
2. *Head Loss Calculation*
   - Use the measured shear stress to simulate the changing pressure loss at the pipe walls
3. *Energy Cost Analysis*
   - Use the pressure loss to determine the energy cost of pumping material at the designed piping distance

Rheometer Results

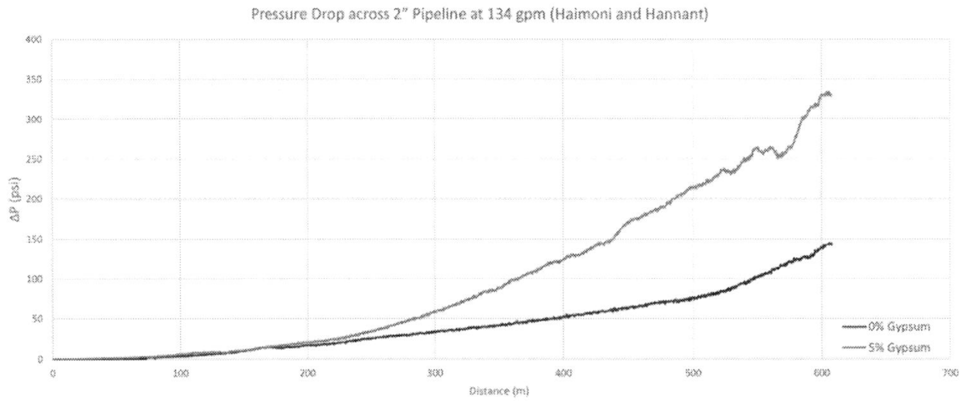

Rheometer Results

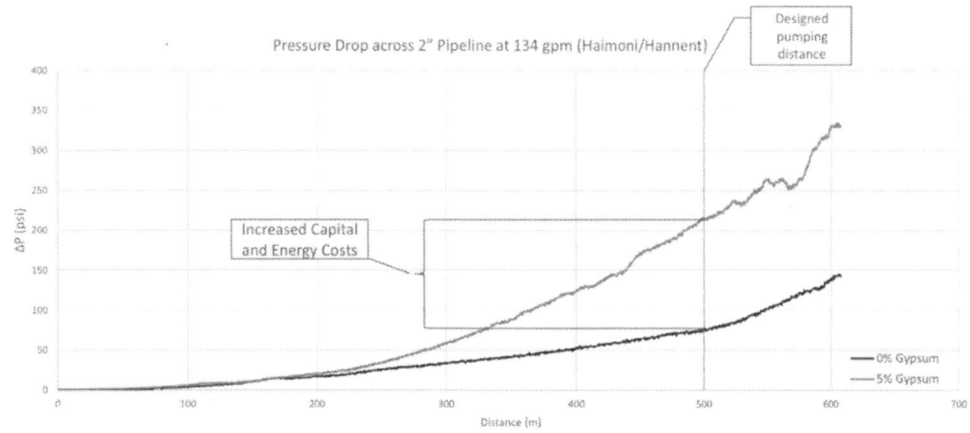

Simulated Case Study: Replacing 5% Gypsum

- Material cost is a constant, but the energy cost increases as time increases
- If pumping beyond the red line, the energy savings outweighs the material savings

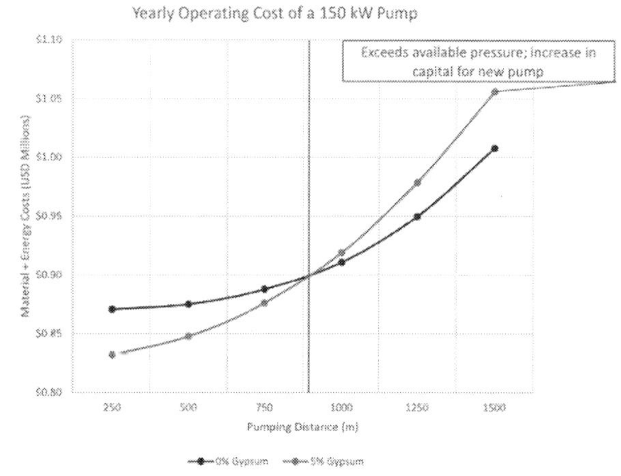

Conclusion

- Next step is confirming these results on a full-scale system
- Simple test that can yield massive amounts of data
- This method can be applied to any variety of time-dependent non-Newtonian fluid

Uncured paste flowing out of mixer

Questions

R&D Research & Development

SOUTHERN COMPANY

511600 Production of Nimesulide Nano-particles via Liquid Anti-solvent Precipitation using Spinning Disc Reactor

Abstract:

This summer I was offered a research fellowship by Indian Academy of Sciences (IAS) to conduct a project on biochemical applications of nanotechnology in Institute of Chemical Technology, Mumbai. According to a study, nearly 40% of the pharmaceutical drugs are poorly water soluble due to which bioavailability of these pharmaceutical drugs is insubstantial and, high dosages of these drugs have to be administered.

My project focused on finding the optimum parameters for producing stable drug nanoparticles in order to increase its dissolution rate. For my study, I chose Nimesulide drug, a non-steroidal anti-inflammatory drug and produced its nanoparticles in a spinning disc reactor using the principles of Liquid-Antisolvent precipitation. The size of the raw drug was found to be 18 microns. The project required a comprehensive knowledge of chemical engineering principles such as transport phenomena, biochemical engineering, reaction engineering and mass transfer. Using these principles, supersaturation of drug was achieved. Spinning disc reactor employed applications of momentum and mass transfer for the production of coated nano-drug. The thickness of boundary layer was also attempted to study.

During my project, I monitored influences of operating parameters such as flow rate of solvent-antisolvent, solvent-antisolvent ratio, nimesulide concentration, disc characteristics and rotating disc speed on the nanoparticles size. At the optimum parameters, particles as small as 190 nm were obtained with a substantial increase in dissolution rate. Samples were analyzed using Dynamic Light Scattering and UV-Vis spectroscopy.

This project is a gateway to finding an efficient method of drug delivery. Using this method, the bioavailability of the drugs could potentially increase which would not only reduce the dosage of the drug administered but also make the drug economically feasible. Production of nano-drug and development of generic pathways for drug delivery could hold the potential cure for epidemic diseases like AIDS, cancer, and tuberculosis.

502651 Design Principles for Graphene-Based Materials to Enhance Supercapacitor Performance

Abstract:

Carbon-based nanomaterials such as graphene and nanotubes are a promising class of electrodes for electrochemical double layer capacitors, due to their high accessible surface area, high electrical conductivity, and tunability through functionalization. A series of recent studies have shown that chemically and mechanically modified graphene electrodes can have improved capacitance, which has been through typically to be associated with the increased double layer capacitance from a high BET surface area or an enhanced electrolyte-electrode interfacial interaction. However, the influence of the electrodeâ€™s quantum capacitance is relatively unknown. In this talk, we will present a computational framework based on density functional theory and classical molecular dynamics to explore the relative contributions of the quantum capacitance of electrode and the electric double layer capacitance to the total interfacial capacitance for various carbon electrode systems in ionic liquids. We will also discuss the impact of the chemical and/or mechanical modifications of graphene-like electrodes on the total capacitance and the design principles to enhance supercapacitor performance.

491694 Property Model-Based Chemcal Substitution and Chemical Formulation Design

Abstract:

Chemical-based products including structured product formulations and single molecule products have proven to be a boon to mankind and have been a significant part of our economies. Our life and the changes around us cannot be imagined without the presence or involvement of chemicals. But like every coin has two sides, some chemicals constituting these products can also be a curse. This is primarily due to the hazardous environment-related properties that some of these chemicals possess and there are many more chemicals which have not been evaluated due to lack of resources for rigorous, experimental-based estimation methods [1]. Hence, there is a dire necessity to identify such chemicals which may be dangerous to the environment, toxic to human health and harmful for our fragile eco-system. Besides, in many cases, their corresponding chemical production processes generate harmful emissions, which also have severe impacts on the environment. The concern about the environment and human health has given rise to the REACH regulation implemented by the European Chemical Agency (ECHA), which compels European chemical companies to stop the use of hazardous substances and replace them with environmentally benign chemicals. Additionally, the decisions taken during chemical product design also have an impact on the process and product performance and are influenced by company strategy, availability of market and government policies [2]. Hence, undoubtedly there is a need to develop a systematic, model-based methodology that can help to find substitutes to existing chemicals in order to improve process economics, operability as well as the sustainability, while still delivering the same or improved product functionality.

It is useful to start with making structured databases by collecting data from knowledge base, followed by the development of the predictive group-contribution property models and then the development of a general methodology for the model-based chemical substitution and chemical product design. The objective here is to quickly and reliably identify the promising candidates through model-based techniques and then to verify and evaluate their performance and applicability through experiments. In this way, the experimental resources are used for verification rather than for an inefficient, trial-and-error search used for practically all chemical products. Besides, when it is desired to come up with alternative substitutes for the undesirable chemicals, the trial an error based approach will have a very large search space. This could be avoided by having predictive models coupled with the desired target properties, making the identification of these substitutes easier. The goal therefore is to investigate comprehensively the uses and properties of the chemicals of concern; develop a systematic framework to identify, compare and select safer alternatives to these including their corresponding manufacturing processes; and finally design safe chemical product formulations or product formulations with improved product performance.

The model-based approach makes use of validated property models to identify the chemicals which need to be substituted, that is, the chemicals that meet the desired physico-chemical properties but not the regulatory (EH&S: environmental, health and safety) properties, and then to generate, evaluate and identify candidates that can replace them. The presentation will discuss the general methodology for chemical substitution, which caters to different problem definitions depending on the reason for substitution. The associated property modeling tools [3] will also be highlighted. A set of new group contribution-based models for a number of useful properties of amino acids will be presented. Through examples on substitution of chemicals from chemical-based products from various sectors namely cosmetics and personal care, pharmaceutical and food, with amino acids (as active ingredient or as additive) will be shown along with other well-known substitution problems. These examples will also highlight the role of property models in chemical substitution and chemical product formulation.

References

1. Hukkerikar, A.S., Kalakul, S., Sarup, B., Young, D.Y, Sin, G, Gani, R, Estimation of Environment-Related Properties of Chemicals for Design of Sustainable Processes: Development of Group-Contribution+ (GC+) Property Models and Uncertainty Analysis.Â *Journal of Chemical Information and Modeling*, 52 (2012): p.2823â"2839

2. Fung, K., Ng, K., Zhang, L. and Gani, R., A Grand Model for Chemical Product Design.Â *Computers and Chemical Engineering*, 91 (2016): p.15-27

3. Gani, R. Computer-Aided Methods and Tools for Chemical Product Design. *Computer Chemical Engineering Research and Design*, 82(11) (2004): p.1494-1504

498103 The Chemical Product Design of Solvents for Post-Combustion CO_2 Capture Processes

Abstract:

The solvents used in post-combustion CO_2 capture processes must satisfy several physical, chemical, operational and economic constraints. They must possess a reaction equilibrium with CO_2 that is significantly affected by temperature. They must possess vapor pressures that allow for easy separations. They must possess heat capacities and enthalpies of vaporization that lower the energy requirements for thermal processing.

We used a chemical product design approach to systematically identify promising candidate solvents for the capture of CO_2 from coal fired power plants. Our chemical product design approach consisted of the following five steps:

1. Compiling and developing accurate structure-property relationships for the basic physical properties needed for performance evaluation. These properties include activity coefficients, pKa, heat capacities, vapor pressures and enthalpies of vaporization.
2. Using a series of structure generation and enumeration algorithms to create a database of candidate solvents. These generated molecular structures include both commonly known solvents as well as novel solvent structures.
3. Using the compiled structure-property relationships to estimate the physical properties of each candidate solvent.
4. Quickly evaluating each candidate solvent using a simplified process-independent model.
5. Detailed evaluation using a process simulator.

We will discuss the details of our approach, the details of the algorithms used and areas in which improvement is needed.

499087 The Effect of Reactor Geometry on Polymer Composition Broadening

Abstract:

The Effect of Reactor Geometry on Polymer Composition Broadening

Pradeep Jain, Ivan Konstantinov, Thomas Karjala and Carlos Villa

The Dow Chemical Company, Freeport, TX 77541

The stochastic nature of polymerization processes results in mixtures of chains that differ from each other in their molecular architecture. Thus, even under constant process conditions across both time and space, one should not expect chains with the same number of repeating units due to the interactions among the various steps of the polymerization mechanism. Other factors also play significant roles, with two of the most important ones being reactor geometry and mixing. Our work focuses on the impact of all these effects on the polymer composition distribution.

Polymer composition is a key quality characteristic. In the production of Linear Low Density Polyethylene (LLDPE), Î ±-olefins such as butene, hexene, and octene introduce small side chains into the ethylene based polymer backbone. These side chains reduce the density and the crystallinity of the final polymer, a desirable characteristic for certain applications. Increasing levels of Î±-olefins can also lead to products with lower molecular weights, since Î±-olefins are typically not as reactive as ethylene towards chain growth. The strong dependency of key product properties on the level of incorporation of Î±-olefins comes at a cost, however, since missing a target by adding too much or too little of these comonomers can lead to products with unexpected properties.

Using polyolefin copolymerization as an example, we rely on computational tools to study the spread in the polymer composition and molecular weight distributions under the influence of the factors mentioned earlier. We focus on process variables such as temperature, relative monomer concentration, as well as catalyst selection, i.e. reactivity ratios, to generate a product with the desired composition distribution. Furthermore, this work uses various geometries and builds on our previous investigation to illustrate these problems and demonstrates that product quality is highly sensitive to what most would consider simple details in polymerization reactor design.

502690 Parameter Estimation of Nonlinear Stochastic Models: Reactivity Ratio Studies in Copolymerization

Abstract:

Mathematical models provide new ways to understand and describe complex systems. There are two key steps in the model development and design, i.e., parameter estimation (or model calibration) and optimal experimental design. Parameter estimation identifies the unknown parameters of the model, which can provide the best fit to a set of available experimental data. In contrast, optimal experimental design aims to devise the dynamic experiments that can provide the maximum information content for model identification, discrimination, and estimation. In this work, we focus on parameter estimation of nonlinear polymerization models using two popular stochastic modelling techniques, i.e., Markov Chain Monte Carlo (MCMC) and generalized polynomial chaos (gPC).

Nonlinearity in models may affect the way that a model is being interpreted, and linearization-based parameter estimation may be biased sometimes. For example, Kalman filters-based methods usually exert a restrictive assumption about the distribution of parameters to obtain desirable estimates. In addition, uncertainty, originating from noisy data, is another main restrictive factor for accurate parameter estimation. It is useful to propagate uncertainty into the model and study its effect on model predictions.

To estimate parameters of a nonlinear model in the presence of uncertainty, the error in variable model (EVM) was recently proposed. This treats each measurement as an unknown true value plus an error term and can handle implicit models. The disadvantage of EVM is that it is only tractable when all the distributions describing variations in the measurement are normal distributed. Another issue with EVM is that it may converge to a local optimum rather than a global optimum for parameter estimation.

As an alternative, Markov Chain Monte Carlo (MCMC) and generalized polynomial chaos (gPC) can overcome the limitations of EVM-based method. Using these techniques, we present a robust and efficient way to calculate parameter estimates and determine parameter uncertainty. The estimation of reactivity ratios from composition data using the Mayo-Lewis and Meyer Lowry models will be used to illustrate the efficiency of the proposed methods. In addition, comparisons will be provided between the MCMC and gPC results and the results obtained from classical regression approaches.

جامعة الملك فهد للبترول والمعادن
King Fahd University of Petroleum & Minerals

17AIChE Annual Meeting, Minneapolis, MN

Processing and Compositional Characterization of Khulays Clay for Drilling Fluid Application

Jimoh K. Adewole,
Abdullah S. Sultan, Lionel T. Fogang, Mohamed A. N. Mahmoud

College of Petroleum Engineering & Geosciences
King Fahd University of Petroleum & Minerals,
Dhahran 31261, Saudi Arabia

Presentation Outlines

- **Introduction to Drilling Fluids**
- **Literature Review**
- **Methodology**
- **Results and Discussions**
- **Conclusion & Recommendations**

KFUPM

17AIChE Annual Meeting, Minneapolis, MN

Introduction

Innovative Drilling Fluid Technologies

Time

Costs

Environmental

Successful completion of oil well drilling operations

Introduction (Cont'd)

Drilling Fluid and its Component

- Drilling fluids are mixture natural and synthetic chemical compounds
- Used as circulating fluid in drilling operations
- Successful completion of wells depends to a large extent on the properties of the drilling fluid used during the operation
- Types
  - ✓ Water-based mud
  - ✓ Oil-based mud
  - ✓ Gas

Introduction (Cont'd)

Functions of Drilling Fluids

- ✓ Transportation of cuttings
- ✓ Cooling and Cleaning of bit
- ✓ Friction Reducer

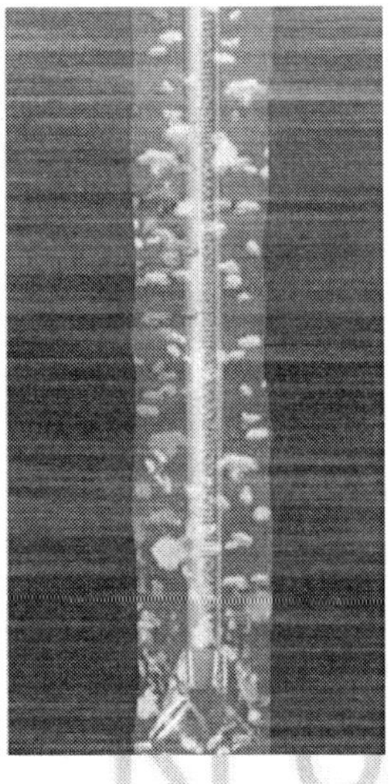

17AIChE Annual Meeting, Minneapolis, MN

Introduction (Cont'd)

Functions of Drilling Fluids

- ✓ Maintaining borehole stability
- ✓ Prevent inflow of fluids from permeable rocks that have been penetrated
- ✓ Formation of thin filter cake to seal pores/other openings

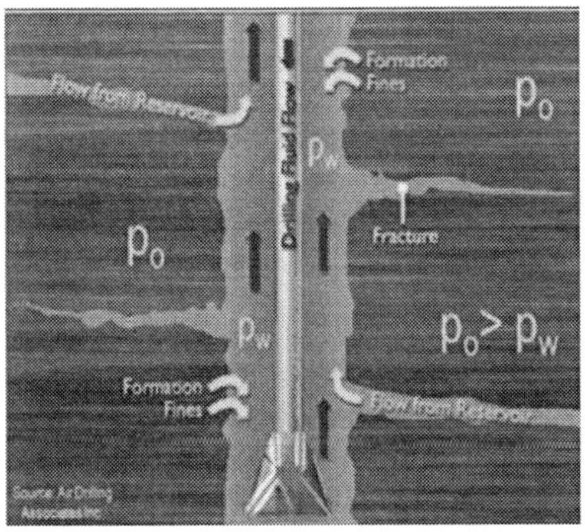

17AIChE Annual Meeting, Minneapolis, MN

Introduction (Cont'd)

Performance Requirements of Drilling Fluids

- Non-damaging/offensive
- No expensive completion methods
- No interfere with productivity of the fluid bearing formation
- No corrosion/excessive wear of drilling equipment

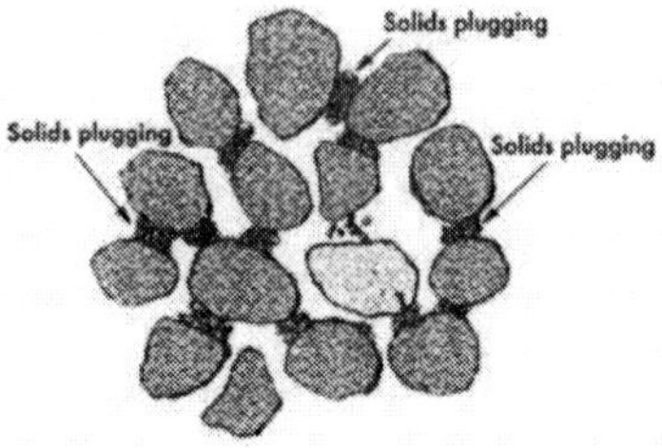

Formation damage caused by solids plugging (petrowiki.org)

Introduction (Cont'd)

Drilling Fluids Selection

- **Cost**
- **Environmental concerns**
- **Sustainable sources of raw materials**
- **Formation/Well Types**
- **Water Requirement**
- **Others**

Research effort in drilling fluid has been directed towards developing cheap materials in order to reduce the cost of drilling

Introduction (Cont'd)

- Raw bentonite are naturally occuring minerals
- Major component is Monmorillonite
- Contains various impurities: clay and non-clay

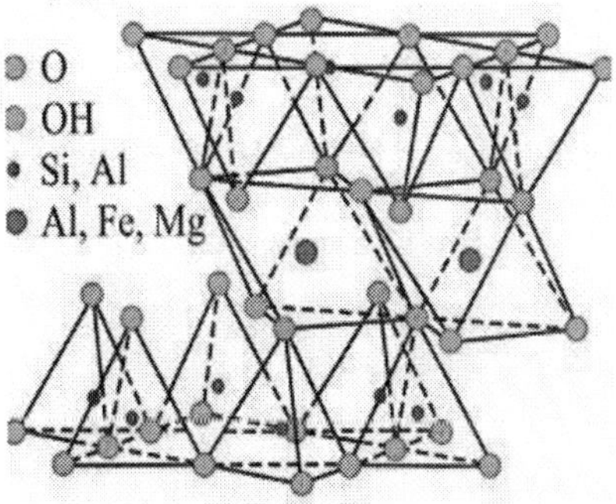

Structure of Monmorillonite
(Poppe et al, 2002)

Site of bentonite deposit

Introduction (Cont'd)

- Khulays clay deposits is 95km north of Jeddah (adjacent to Makkah-Madinah road), Saudi Arabia
- The largest clay deposit in the Kingdom
- Area of 800m length by 600m width and 200 – 300m depth
- Estimated to contain 3million tons of raw clay (Magzoub et al, 2017)
- Khulays Clay (Bentonite) (Al-Shafei et al, 2017)
  - ✓ Na montmorillonite 58 -90 %
  - ✓ Fedspar 5%
  - ✓ Quartz-SiO_2 15 - 25%
  - ✓ Kaolinite 8-10%
  - ✓ Halite 2%
  - ✓ Illite 1%

Project Objectives

Develop a cost effective technique for processing locally sourced raw materials from Khulays clay deposits for application in drilling operation

- ✓ Conduct a preliminary laboratory scale research
- ✓ Establish a systematic sets of procedure for processing and purification of the clay
- ✓ Provide some data to aid the applied research and product development moves into larger laboratories, pilot and production plants

KFUPM

17AIChE
Annual Meeting, Minneapolis, MN

Khulays Clay for Drilling: Literature Review

- The possibility of using local bentonite clay minerals for drilling fluids formulations (Al-Homadhi, 2007)
- Used:
  - ✓ 5 - 15wt% Khulay Bentonite
  - ✓ Soda Ash and Drispac
- Results showed:
  - ✓ Yield value of 85 bbl/ton (compared to 95 bbl/ton imported bentonite)
  - ✓ Yield can increase up to 90bbl/ton by 0.5 % Drispac
  - ✓ Identical viscosity and filtration loss values
  - ✓ The cost of the enhanced Khulays Bentonite can be less by about 46% of than that of imported bentonite

KFUPM

© King Fahd University of Petroleum & Minerals | Dhahran, 31261 | Saudi Arabia | +966 (3) 860 0000

17AIChE
Annual Meeting, Minneapolis, MN

Khulays Clay for Drilling: Literature Review

- Magzoub et al, 2017 disclosed a method for producing sodium bentonite from Khulays clay
- The resulting drilling fluid has the following properties:

Properties	Commercial Bentonite	*Treated Khulays Bentonite
Filtration (ml)	15	28.4
Apparent Viscosity (cp)	11.5	20
Yield Point (lb/100ft2)	7	28
Na/Ca	2.896	2.950

*Treated with 1g Soda Ash & 24h of heating and stirring

Khulays Clay for Drilling: Literature Review

- Al-Shafei et al (2017) disclosed a method for modifying Khulays bentonite for application as drilling fluid
- Raw water for removing contaminant form the bentonites
- Additives
  - ✓ Polyanionic Cellulose Polymer (PAC): 1 – 5%
  - ✓ Soda Ash: 1 – 3%
  - ✓ Sodium Carboxylmethyl Cellulose (CMC): 5 – 7.5%

Properties	Commercial Bentonite	*Treated Khulays Bentonite
Filtration (ml)	15	10
Plastic Viscosity (cp)	27.5	24.5
Yield Point (lb/100ft2)	16	19.5

Methodology

Sample Collection and Processing

- ✓ Samples were extracted from outcrop of Khulays deposit
- ✓ Eight samples of Khulays clay collected from various parts of the outcrop were analyzed
- ✓ Analytical tests were performed to identify the various types of clays that are contained in each of the raw samples
- ✓ Drying
  - Drying Time 16hrs
  - Drying Temp 110°C
  - Moisture removed ~ 12wt%
- ✓ Grinding for various duration of time
- ✓ Sieving and particle size analysis

Methodology

Analytical Characterization

- Compositional Analysis
  - ✓ X-ray powder diffraction (XRD)
  - ✓ scanning electron microscope (SEM)
  - ✓ Energy Dispersive Spectroscopy (EDS)

Results and Discussions

XRD Analysis

Q = Quartz
C = calcite,
b = Biotite
A = Feldspar (Albite)
M = Montmorilonite
K = Kaolinite

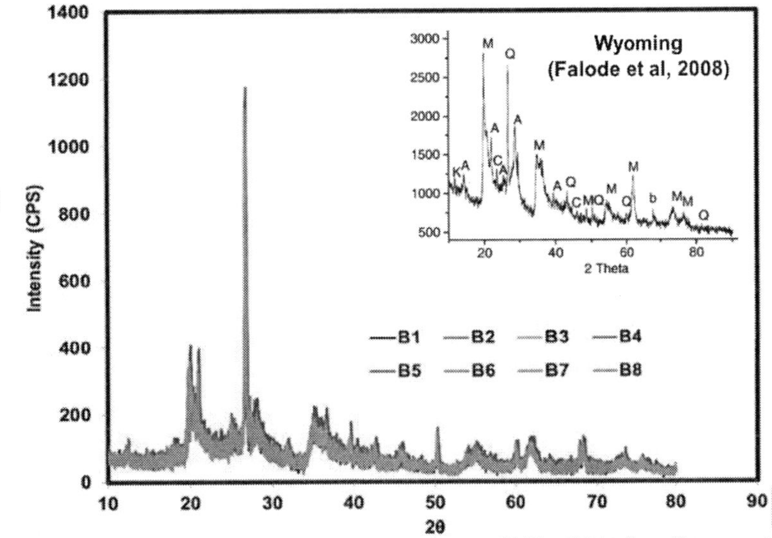

Fig. XRD Spectrum of samples from various locations compared with Wyoming Bentonite

Results and Discussions

Morphological images from SEM

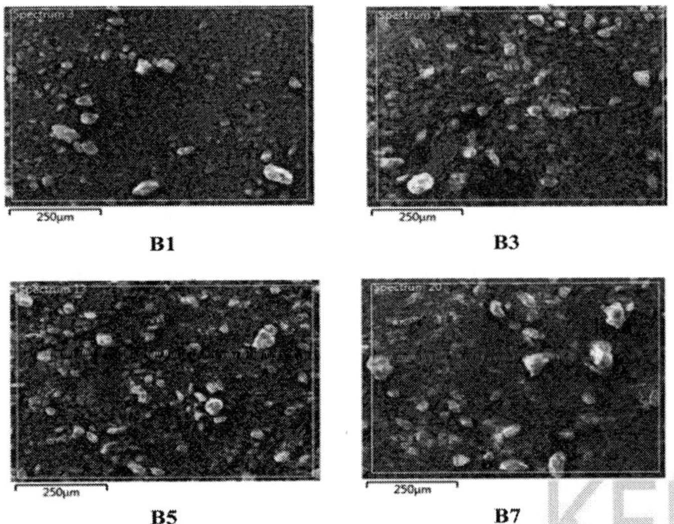

Fig. SEM Images of selected samples from various locations

Results and Discussions

Elemental Composition by EDS

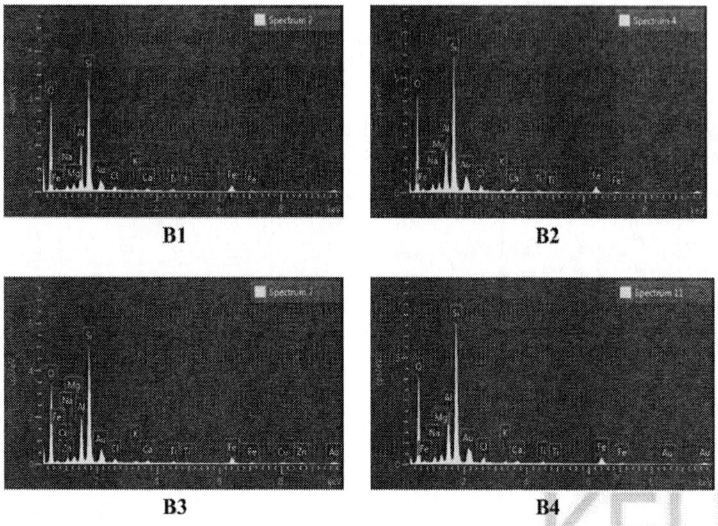

Fig. EDS Spectrum of the element of sample from various locations

Results and Discussions

Elemental Composition

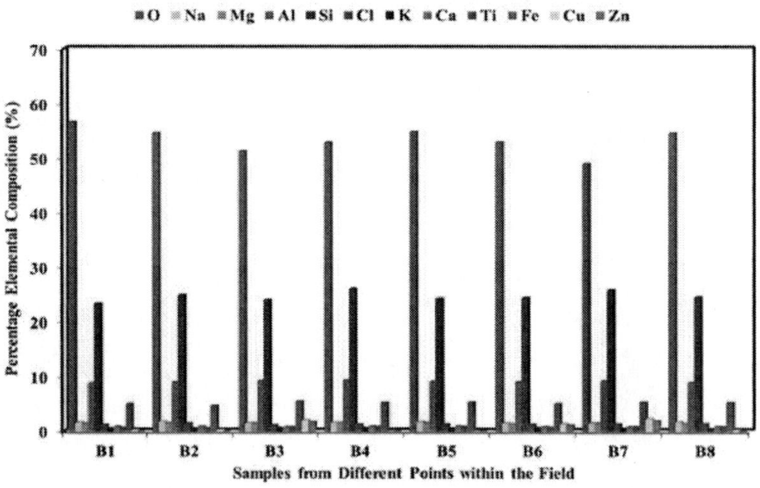

Fig. Elemental compositions of sample from various locations

Results and Discussions

Na^+/Ca^{2+} Ratio

Element	B1	B2	B3	B4	B5	B6	B7	B8
Na	1.25	1.34	1.02	1.38	1.41	1.46	1.10	1.49
Ca	0.85	0.82	0.77	0.82	0.93	0.84	0.83	0.82
Na/Ca	1.47	1.63	1.32	1.68	1.68	1.74	1.33	1.82

All samples seem to be more of Na based Bentonite than Ca

KFUPM

17AIChE
Annual Meeting, Minneapolis, MN

Results and Discussions

Sample properties based on locations

Filtration and rheological properties of samples from various locations

Samples	Density	Fluid Loss (ml)	Mud Thickness (in)	Apparent Viscosity (cp)
1	8.7	108	0.174	2.5
2	8.7	144	0.228	2.5
3	8.7	80	0.126	2.5
4	8.7	124	0.200	2.5
5	8.7	108	0.197	2.5
6	8.7	147	0.263	2.5
7	8.7	125	0.235	2.5
8	8.7	128	0.238	2.5

- ✓ **Performance properties are affected by other factors than Na/Ca ratio**
- ✓ **Subsequent experiments utilized mixture of samples**

© King Fahd University of Petroleum & Minerals | Dhahran, 31261 | Saudi Arabia | +966 (3) 860 0000

17AIChE
Annual Meeting, Minneapolis, MN

Results and Discussions

Sample Processing and Purification- Grinding

Effect of Sample Quantity (Raw Samples)

Grinding Time (hr)	Sample Quantity (g)	%wt ≥ 90μm
2	745	12.05
1	664.5	10.59
1	233.6	29.83

Effect of Sample Quantity (Purified Samples)

Sample	Sample Quantity (g)	Weight Retained (g)	%wt ≥ 75μm
I	86.50	1.90	2.20
II	194.40	62.10	31.94
III	219.60	38.90	11.71

✓ **Grinding time for purified samples was 1hr**

Results and Discussions

Sample Processing and Purification

Sample recovery and use of water

Samples	Initial Weight (g)	Recovered	Bottom	Recovered (%wt)	Time (days)	Water Used (Lt)
≤ 75um	50.29	8.71	41.58	17.32	5	15
≤ 75um	50.00	34.87	13.97	69.74	15	21
90um ≤ 125um	50.01	28.65	20.92	57.28	15	21
125um ≤ 150um	50.00	28.27	21.62	56.54	15	21

The use of large amount of water necessitates the need for water recycling

- **Subsequent processing done for:**
  - ✓ **15 days**
  - ✓ **Recycled water**

Results and Discussions

Multistage Purification

- Multistage purification was investigated with 100g clay in 2000ml of water.
- Three different stages were investigated: 1, 2 and 4 stage recovery

Cycles of Recovery

- The numbers of recovery cycles for each of the stages were also investigated
- This was done by measuring the height of the settled bentonite in solution after recovery

Results and Discussions

Multistage Purification & Recovery Cycles

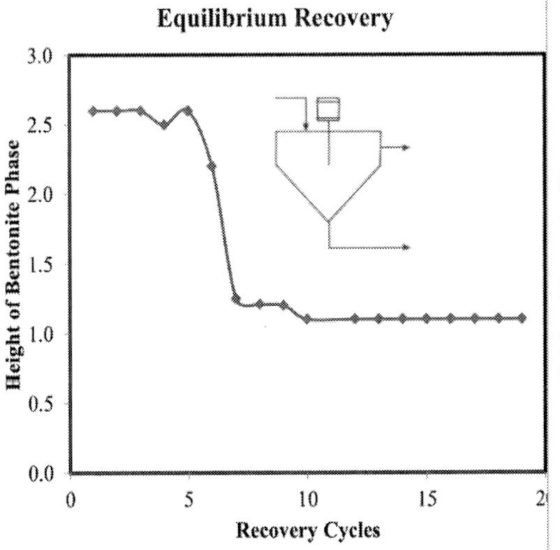

Fig. Single - Stage

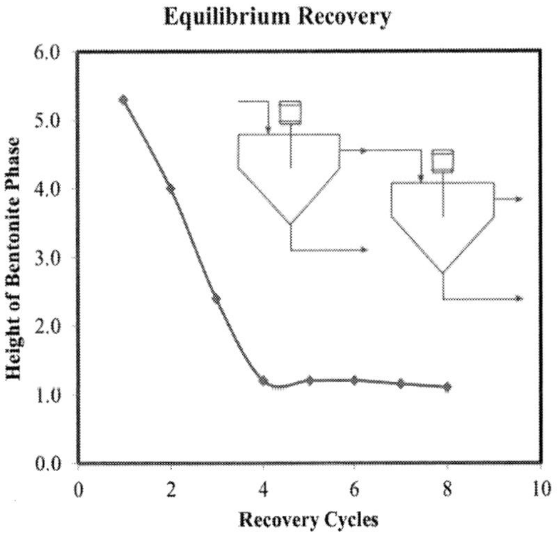

Fig. Two - Stage

© King Fahd University of Petroleum & Minerals | Dhahran, 31261 | Saudi Arabia | +966 (3) 860 0000

17AIChE Annual Meeting, Minneapolis, MN

Results and Discussions

Multistage Purification

Table. Performance Properties of samples from the multistage processing

Sample	Density (lb/gal)	Filtrate Collected (cc)	Cake Thickness (in)	Viscosity @ 600 (cp)
Commercial	8.5	38.2	0.0563	7.00
Khulays-Stg I	8.5	220	0.106	4.00
Khulays-Stg 2	8.5	129	0.1185	4.00
Khulays-Stg 4	8.5	238	0.1197	4.00

- ✓ **Similar performances with respect to rheological properties**
- ✓ **Two-stage recovery looks better in terms of filtration properties**

Results and Discussions

Filtration Properties o
Khulays and Commerci
Bentonite

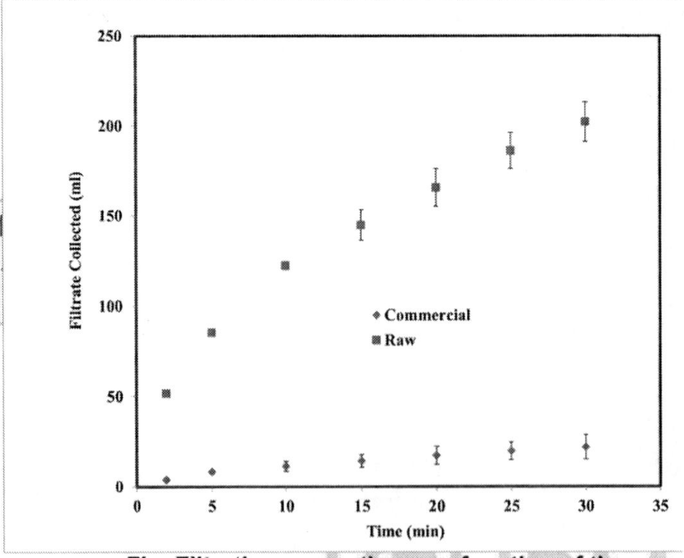

Fig. Filtration properties as a function of time

Conclusion and Recommendations

- Product development involves a blend of science and business
- The business manager and manufacturers must be convinced that Khulays bentonite will be profitable and can be manufactured cost-effectively
- Sample compositions varied very slightly from one point to the other within the field
- Conduct more experiments to ensure the main features and functions of the bentonite formulation are retained as other properties are changed
- Working with process engineers to scale up processes in order to make the bentonite in larger quantities for commercialization
- Identification acceptable ranges for purity of the raw and final bentonite material
- Develop a robust Design of Experiments methodology to learn as much as possible in the fewest number of experiments

17AIChE
Annual Meeting, Minneapolis, MN

Acknowledgement

- **College of Petroleum Engineering and Geosciences, King Fahd University Petroleum Minerals, Saudi Arabia**
- **Mr Musa O. Najimu, ChE, KFUPM**
- **Mr Kazeem Muhammad, ChE, KFUPM**
- **Hatim Dafalla, CER, KFUPM**
- **Abdulrashid I. Mohamed, CER, KFUPM**

KFUPM

© King Fahd University of Petroleum & Minerals | Dhahran, 31261 | Saudi Arabia | +966 (3) 860 0000

17AIChE
Annual Meeting, Minneapolis, MN

Acknowledgement

This presentation should be cited as:

Adewole et al (2017)

Referenced as:

Jimoh K. Adewole, Abdullah S. Sultan, Lionel T. Fogang, and Mohamed Mahmoud (2017), Characterization and Processing of Khulays Clay for Drilling Fluid Application, AIChE Annual Meeting in Minneapolis, MN, USA, Oct 29 - Nov 3, 2017

KFUPM

17AIChE Annual Meeting, Minneapolis, MN

494479 Multi-Objective Optimization for Optimal ORC(organic Rankine cycle) Design Considering Inherent Risk, Exergy Efficiency

Abstract:

Inherent safety is technique for designing safer process, which minimize risk of process. This technique give insight about process risk and strategies for reduction of risk in early stages of process design. Most of all, if there are some risky points in the process design, the design can be changed cost-effectively due to characteristics in early stages of design. However, research for inherent safety still need complicate progress and depends on subjective judgment due to lack of a standard measurement method. In addition, risk reduction is applied to complete process design in specific design stages. After risk reduction, if severe decrease of energy efficiency happen, improvement of design is necessary again. Then inherent safety technique will be applied again, and these steps may be repeated. Therefore, decision method is needed, which decides between risk and energy efficiency for proper design. Organic Rankine cycle(ORC), which use organic working fluid, is sustainable energy utilization technology by converting low temperature(low quality) heat to electricity. Exergy efficiency, which means available work or availability, is the best parameter to measure performance of ORC. Exergy efficiency of ORC depends on a sting of variables, which operating pressure, kinds of heat source and so on. Most of them, kinds of working fluids and their composition are effective parameters. And its structure and working condition are effective parameters as well. In this study, inherent safety technique, which contains decision method between risk and efficiency, is proposed and applied to ORC process. The ORC uses liquefied natural gas(LNG) as cold exergy source, which is not utilized in general. Proposed method is considering risk(or safety) and efficiency(or process) simultaneously by solving optimization problem. As a result, multi-objective optimization method is used for optimal design considering safety aspect and process aspect. Optimization variables include working conditions(pressure, temperature), working fluid conditions(composition) and ORC structure ORC. In conclusion, Pareto-optimal solutions are obtained and best optimal solution is selected by decision technique.

496334 A Study on the Explosion Characteristics in Blast Test Facility By Flammable Gas

Abstract:

Facilities to reduce the extent of damage caused by accidents such as flammable gas leaks, fire, and explosions in offshore plants are fire walls and explosive walls. Of these, The design of the explosion wall is performed through structural analysis by selecting the load derived from explosion experiment or simulation. However, this method can be applied only to the simple design of explosion-proof walls.

In this study, the congestion of the internal structure in the limited space and the explosion characteristics according to the combustible gas were confirmed by CFD simulation. The overpressure measured at the test facility when the congestion is 0.3 is large and these results can be referred to when designing the equipment for the blast test facility.

Experimental Safety Plan for Safety Management in Chemical Engineering Research

Juanita Miller, David Rockstraw, Martha Mitchell, Derrik Wootton

Abstract:

Academic and industrial research is fast paced and cutting edge which presents many safety challenges. The nature of research invites shortcuts and improvised techniques that can lead to hazardous conditions. Documentation on processes are ever changing and frequently inaccurate. Decisions by young and inexperienced workers can lead to poor judgment and planning. Incidents from universities and research labs have resulted in several injuries and a few fatalities. Countless near-misses have undoubtedly occurred without any statistical data or documentation collection. Some of the best know are the accidents at UCLA in 2008, Texas Tech in 2010 and most recently University of Hawaii in 2016.

We will discuss a rigorous approach to documentation and training as well as methods for analyzing the hazards of experiments and surroundings to benefit research safety. This system of research experiment management using the Experimental Safety Plan (ESP) approach challenges the stakeholders to consider hazards of their experimental work, consequences of a hazardous situation and methods to mitigate the risks. We will also discuss the history of laboratory and research accidents, the benefits to safety, quality and research documentation, and methodology of Experimental Safety Plan used at NMSU's Chemical Engineering Department. This method will also be compared with guidelines given by the Department of Labor in 29 CFR 1910.1410 Occupational Exposure to Hazardous Chemicals in Laboratories.

Keywords: Safety, Laboratory, Education

Corresponding author: College of Engineering, New Mexico State University, P.O. Box 30001 MSC3449, Las Cruces, NM 88003, USA. Tel: +52 575 646 1292; fax:+52 575 646 3549 E-mail: miljgh@nmsu.edu

Background:

What if one of your researchers was severely injured by an explosion caused by static electricity igniting a mixture of hydrogen, oxygen and carbon dioxide in a compressed gas tank? A biofuels experiment had been run successfully for a couple of years using fixed bottles hydrogen, oxygen and carbon dioxide gas that were mixed in situ at the reactor. The professor had a better idea to build a mixed gas tank out of a 13 gallon compressed air tank that could be filled with the needed gas ratios on site. The professor and researcher designed and constructed the gas system complete with tank, valves, fittings and digital pressure gauge. They decided to keep the tank portable and carry it to the fill location and to reactor as needed. The design of the mixed gas system was a process change but it no one evaluated it for potential safety hazards. Earlier in the week the researcher experienced a small internal explosion with a similar one gallon tank assembly when she pressed the OFF button on the pressure gauge. This was a near miss and should have alerted the researcher and professor of a serious problem. During that week the researcher ran the experiment again with the larger mixed gas tank assembly with disastrous consequences. When she disconnected the hoses used to fill the tank and pressed the OFF button on its pressure gauge, an explosion occurred. (*1*)

This is a true story from the University of Hawaii in March 2016. The researcher suffered injuries including loss of her right arm above her elbow, abrasions to her cornea, burns on her face, and nerve damage to her ears with resulting loss of high frequency hearing. And a civil lawsuit has been filed on her behalf, claiming that her supervisors "failed and refused to warn of the dangers or to provide appropriate safety training."(*2*) Damage to the lab and building were between 0.5 and 1.0 million dollars. And the university was fined $115.500 from Hawaii OSHA. The agency labeled all 15 violations as "serious" and assessed the maximum state penalty of $7,700 to each (*3*). A subsequent investigation showed that the root cause of the explosion was likely static electricity and not the digital pressure gauge as the Honolulu Fire Department originally determined.(*4*)

An incident like this becomes part of the legacy of a university. When someone searches for information on the school or an academic department, a story like this will appear in the list of possible matches and persists as bad advertising. Not only do potential students see this but grant organizations, donors and politicians. We need to learn from incidents like this and build a culture of safety that encourage Principal Investigators (PI) and researchers to recognize hazards and question if experiments can happen in a safer manner.

Several items were highlighted as deficiencies in the University of Hawaii accident. Among them lack of safety review of the experimental design, personal protective equipment (PPE), fire and explosion hazards, and emergency procedures. Also sited was lack of training, and inspection of laboratories and procedures.

Some wise people at NMSU recognized that oversight of Chemical Engineering research was needed starting in 2007. Researchers were ordering chemicals and conducting experiments with limited oversight. Experimental samples were collecting in bins and drawers and remaining long after a researcher graduated. At some point, when no owner was found, unclaimed samples had to be disposed. Some without proper labeling required costly testing to determine their composition prior to disposal. Process hazards were going unrecognized due to lack of training and researcher awareness of a safety culture. NMSU's ESP process was born out of a recognition of safety hazards and potential risks. It has gone through several levels of development. It has taken the guidance of two dedicated department heads and support from the university EHS department to achieve this professional document. And it continues to evolve to better ensure safety but not impede research activities.

Approach:

Much of the framework for the ESP follows guidance given in a 2013 publication of the American Chemical Society, titled "Identifying and Evaluating Hazards in Research Laboratories". Chapter 3, "Hazards Identification and Evaluation", lays out many of the components incorporated into the ESP, such as defining the scope of work, hazard identification, selection of hazard controls and performing work within controls. In Chapter 4, "Establishing Roles and Responsibilities", this document discusses all party's obligation to participate in the laboratory safety culture. The premise of the ESP process is for the researcher to write and explain the planned experiments and the PI to review it. This ensures that all the major parties are engaged with safety for their experiments. Chapter 5 gives suggested methods of hazard analysis. The ESP process has been evolving to include an industrial hygiene analysis of chemicals used, and various hazard analysis techniques including a technical review, 'What If' analysis, FMEA and HAZOP. Chapter 6 of this document addresses change control. The ESP process has been evolving to include unique tracking numbers and revision control for each experiment. This also allows ESPs to be transferred between researchers. (*5*)

Much of the spirit of OSHA 1910.1450 can be seen in the ESP process with the companion university chemical hygiene plan and effective communication of employee information, training and hazard identification into this process. The Appendix A of this regulation discusses the culture of safety which the ESP process seeks to encompass.

Experimental Safety Plan Format Details:

The ESP begins with basic information on the experiment, title, names of PI and researchers, and defines the location for the work. This also contains an ESP number, revision number and signature block. A revision block with notes on changes from each version allows for future reference and tracking. This also allows a given ESP to be transferred or handed-off between researchers when graduations occur or between PI's if their research efforts change course. With several dozen active ESPs at any given time, a tracking system was critical to ensure the information was maintained in an orderly fashion. Theoretically a given ESP could exist for several years as the research continues, providing valuable information to a PI on the progress of the research project.

A formal process for evaluation of experiments, along with chemicals used, waste generated and site related hazards was established. This process begins with the researcher who must explain, in writing, details of the experimental work that is planned. The Principal Investigator must review this document for completeness and accuracy. Then it is submitted to the College Safety Specialist and the Department Head for review.-Any changes to an approved document are made through a revision control process with highlights for differences between the old and new versions.

Attachment 1 - Experimental Scope: According to the American Chemical Society, "an important, but often missed, preliminary step…is identification of the task or group of tasks to be evaluated. Without this, the effectiveness of every subsequent stop in the process can be compromised." It further goes on the say that "organizations often find merit in establishing guidelines around scope determination that fit well with the type of research activities being performed." Thus the ESP scope is intended to give a thorough overview of the work to be performed by answering the following questions:

- Explain why the work is being performed, the goal(s) of the experimental program - If this is an update/revision of previous ESP describe all changes
- Provide the stoichiometry of any chemical reactions and their heats of reaction
- Demonstrate the inherent thermal safety of your experiment through calculation or through the use of accelerating rate calorimetry data.
  Include a complete list of all chemicals (reactants and products) involved in the work.
- Include a complete list of all equipment (e.g. autoclave, centrifuge, pump, heat bath etc.) involved in this work

Attachment 2 – Drawing of the laboratory or pilot area: The researcher is asked to provide a detailed drawing of the laboratory or pilot area in which the work will be performed including locations of the experimental equipment, safety equipment (including eyewash stations and safety showers, fire extinguishers, first aid kids, noting the date(s) of last inspection of these safety devices), Safety Data Sheet (SDS) compilation, chemical storage, and evacuation route. This information allows for a review of hazards within the area including conflicts and compatibility between other researcher's experiments.

Attachment 3 – Normal Operations, Startup and Shutdown Procedures: The researcher is required to provide a step-wise procedure that describes in detail how the work will be performed. The procedure should begin and end with the equipment in the normal idle (inoperative) state. Answers to the following questions determine the nature of hazards recognized by the researcher:

- Include a statement of the required PPE at the beginning of the procedure, and at every location in the procedure where the PPE requirements change.
- Indicate where hazardous feedstock chemicals will be stored, how they will be transported to the location of the experimental work, how they will be transferred from the storage vessel into the experimental apparatus, and how they will be returned to storage.
- Take into account those items for which you indicate "yes" on the NMSU Lab Hazard Assessment Checklist.

Attachment 4 – Emergency Shutdown Procedure: The researcher is required to provide a step-wise procedure that describes how the equipment will be brought to a safe state in the event of an emergency. The description should include a detailed explanation of how to attend to potential medical emergencies that may result.

Attachment 5 – Waste Management Procedure: The researcher is required to prepare a Waste Management Procedure that provides the exact nature and estimated volumes of all wastes to be generated in performing these experiments. The EHS department reviews the chemicals for compatibility and recommends segregation and storage as required.

Attachment 6 – Hazard Identification and Mitigation: The researcher is required to identify all high hazards associated with the experiment and describe proposed engineering, administrative and personal protection controls to mitigate them. The analysis must consider the following:

- All sources of energy (electric, chemical, hydraulics, mechanical, compressed gases)
- Extreme conditions of pressure or temperature (from flame or steam to cryogenics)
- Chemical storage
- Housekeeping
- Fire potential
- Biological hazards
- Substances that are highly reactive, radioactive, highly flammable, pyrophoric, highly
  toxic, mutagenic, teratogenic, carcinogenic, or have very low exposure limits, high voltage, high RF, X-ray, laser (class 3b or 4),
- High temperatures or high pressure or pressurizing vessels

And for each high hazard provide the following information:

- Description of the high hazard;
- Operational and engineering controls that will be used (based on identified industry
  best-practices used in addressing this safety hazard);
- Required PPE (beyond minimum) when this HIGH hazard is present; and
- Special training (beyond minimum) that is necessary.

Attachment 7 – Safety Data Sheets (SDS) for all chemicals used/generated in experiment: The researcher is required to provide a complete set of safety data sheets for any chemicals used in the experiment. Any chemicals added by revisions must have the safety data sheets provided for it as well. This provides a mechanism for chemical inventory control to ensure that the researcher is only using and ordering chemicals that have been approved through the ESP process.

Administration of the ESP Process:

Technical Review: Once the researcher has written the ESP document and the PI has completed their reviewed, it is routed to the College Safety Specialist, Department Head and Laboratory Manager for technical review. A determination about the need for and depth of a hazard analysis is made. An industrial hygiene evaluation for hazardous chemicals is completed. On many occasions the document is sent back to the researcher with recommended and required changes to incorporate. Guidance in the understanding of potential hazards and mitigations is needed for most researchers since they lack experience and perspective on potential issues that may exist. Hence mentoring and feedback by technical staff are key elements of this process to ensure that the researcher learns not just what to do but why and when to alert others of problems.

The hazard analysis can take several forms from a simple What If to a complex FMEA or HAZOP depending on the complexity of the situation. It is very important that the researcher be involve in performing the analysis and any new researchers be briefed on the items discussed.

Educating researchers of what constitutes change can be challenging. The obvious things are new chemicals for an experiment. But things like changes to piping materials and routing, or product storage vessels can have large safety impacts. As we saw in the first example, a different type of pressure vessel along with static electricity contributed to a very bad accident. Changes should be evaluated with consideration of the original experiment conditions and appended to any hazard analysis.

Conclusions:

This ESP process has undergone several revisions and we continue to find ways to improve the concept. Without a doubt, educating researchers and PI's about the safety culture and how to identify hazards had improved the safety of chemical engineering research at NMSU. In the beginning, the researchers and PI's were resistant to these methods and insisted that this was too much effort. However, they have seen benefits to the research efforts by having documentation for experimental work that can be incorporated into reports, evaluations and thesis papers. By placing the burden of writing the ESP on the researcher, it forces them to take ownership for their experiments as well as their safety. And having an active technical review process allows everyone to contribute knowledge about hazards making it an overall learning process.

References:

1. City and County of Honolulu, Honolulu Fire Department, Fire Investigation Report Number 2016HFD0015761, March 30, 2016
2. http://khon2.com/2017/01/20/researcher-injured-in-university-of-hawaii-lab-explosion-files-lawsuit/ Researcher injured in University of Hawaii lab explosion files lawsuit, January 20, 2017
3. http://cen.acs.org/articles/94/web/2016/09/University-Hawaii-fined-115500-lab.html , University of Hawaii fined $115,500 for lab explosion, September 23, 2016
4. http://cen.acs.org/articles/94/i28/University-Hawaii-lab-explosion-likely.html , University of Hawaii lab explosion likely originated in electrostatic discharge, July 11, 2016
5. Identifying and Evaluating Hazards in Research Laboratories, American Chemical Society, 2013
6. Occupational Safety and Health Administration, OSHA 1910.1450

501757 Risk Mitigation in Technology Commercialization: Responsibilities of the Owner, Designer and EPC Contractor

Abstract:

Risk Mitigation in Technology Commercialization: Responsibilities of the Owner, Designer and EPC Contractor.

A crucial element of front-end project planning is creating a project risk profile that allows for mitigation of risks that materialize during project performance. One of the first steps is selecting the technology which will provide the basis for the Process Design Package. Herein lies a challenge as too often the technology selection is primarily based on financial metrics to provide the best project payback, while a full risk analysis of available technologies could provide a set of alternatives which insure the project progress is not impeded by inability to properly address potential technology performance and contracting risks which arise later in the project.

These challenges are even more prevalent in new technology commercialization. Each party has a different level of understanding of the underlying risks and varying abilities to manage such. By working together to identify potential risks up front and sharing the responsibility of managing the risks, the parties are more likely to enjoy both the successful commercialization of technology and the on-time and on-budget project completion.

The Designer must provide proper development of a Technology Package with an extended PDP and a Design Team which can effectively support the Contractor and the Owner through the engineering, construction and commissioning. The Contractor team must be experienced in derisking new technology implementation and be able to integrate the capable Design Team into the Contractorâ€™s performance. The Owner must provide leadership during the development of the FEED, guiding the team with a clear understanding of the project goals and risk tolerance.

The presentation will include lessons learned from development of a new technology for production of specialty chemicals through implementation of the technologies at multiple plants over a six-year period.Â We will look at an example of how the process design firm mitigated risks on new technology commercialization in a developing industry in order to satisfy Owners, Contractors and management.

503415 A Datacentric Approach to Develop Process Design Packages Destined for Mega Projects

Abstract:

Mega projects often face cost overruns and schedule delays due to their scale and complexity. In order to reduce these associated risks, the PDP Engineering team at SABIC is developing a systems based approach to develop Process Design Packages (PDPs) for processing technologies destined for implementation for mega projects. Process technologies for implementation within these projects are developed through the pipeline of research and development programs, in addition to mature technologies that have been installed in industry. The PDP usually developed in the feasibility phase of a project, substantially increases the technical definition of a project. The critical deliverables in the PDP are the process flow diagrams (PFDs), piping and instrumentation diagrams (PIDs) and equipment datasheets. These deliverables require up to 60% of the total man-hours required to develop the PDP. The systematic data-centric approach to develop primarily these deliverables enables efficiency and standardization. By developing a workflow and templates around the database, the risks associated with mega projects are intended to be managed. Furthermore data available within the database becomes available for: other deliverables within the PDP; later project phases; and reuse in future projects.

Mid-Atlantic Technology, Research & Innovation Center

MATRIC

Market-Driven Innovation

Transfer of Critical Technology Learning and Unknowns to the Engineering Design Package

CHEMICAL, ENERGY & ENVIRONMENTAL TECHNOLOGIES

TECHNICAL ENGINEERING

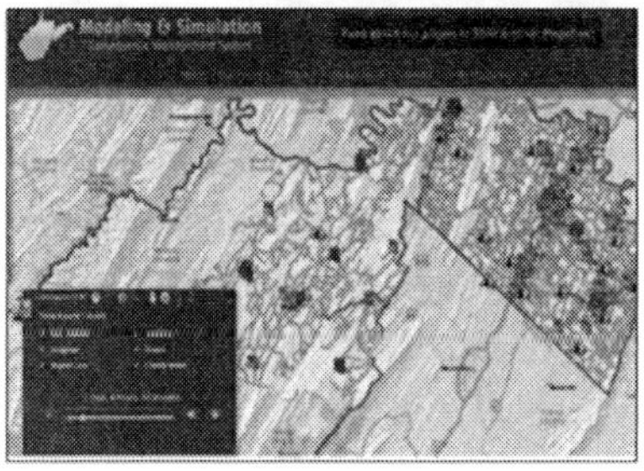

ADVANCED SOFTWARE TECHNOLOGIES

- Confidential -

OVERVIEW

- Technology Development and Transfer Best Practices
- Components of Technology Transfer
- Understanding Technology Uncertainties
- Addressing Technology Uncertainties in the Design, with Examples
- Wrap-up

2

MATRIC

TECHNOLOGY DEVELOPMENT BEST PRACTICES

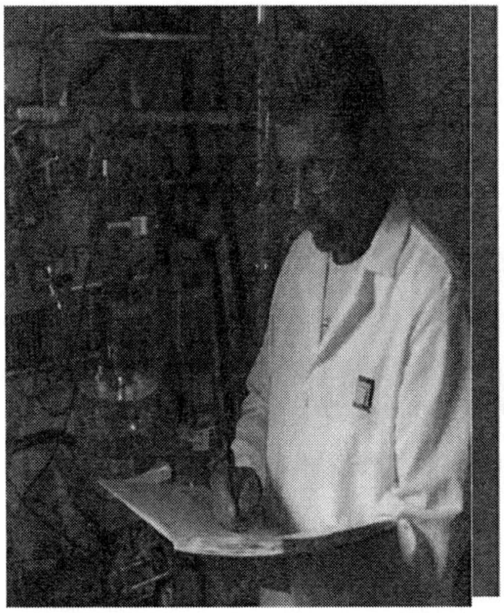

- Multi-functional Project Teams
- Project Staffing Continuity
- Business Engagement in Technology Development at all Stages
- Corporate Culture
- Project Risks Understood and Mitigations In-Place where Possible
- Project Documentation Available and as Complete as Possible

3

MATRIC

TECH TRANSFER BEST PRACTICES

- Use Proven Work Practices
- Engage Multi-functional Teams
  - Focused on Multi-functional teamwork
  - Team Continuity through Implementation
- Document the Tech Transfer in a thorough, organized Definition of Technology

4

MATRIC

Technology Transfer Process

FEL-1	FEL-2	FEL-3
Develop cost effective process design(s), including chemistry, separations, physical properties, and economics. Identify technology unknowns and demonstrate technology feasibility sufficient to justify moving to next phase.	Evaluate process design(s), economics, and operations to select the optimal process. Technology unknowns are identified and resolved.	Complete Basic Facility Scope Package to define process technology for Detailed Engineering, Procurement, and Construction.

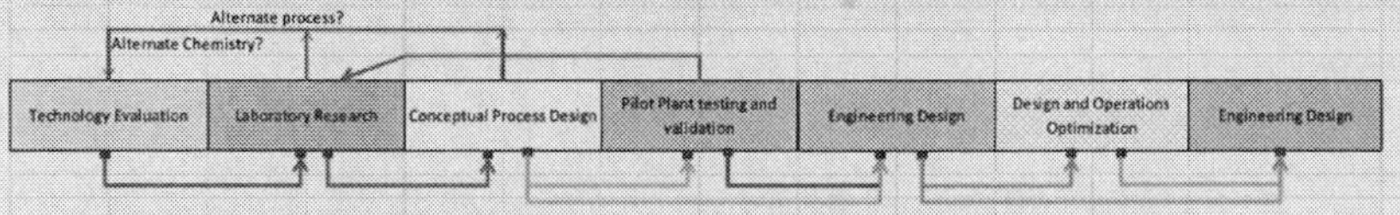

Deliverables	Technology Validation Report that includes:	Technology Studies Report that includes:	Preliminary Definition of Technology that includes:	Definition of Technology that includes:	Preliminary Process Package (FEL-2) that includes:	Process Improvement Assessment Report that includes:	Basic Facility Scope Package (FEL-3) that includes:
	• Economic model describing technology opportunity and needs • Technology Description • Process Models • Experimental methods • Experimental results • Experimental interpretation • Description of prioritized technology and economic uncertainties • Proposed plan to address uncertainties	• Technology objective • Experimental methods • Experimental results • Experimental interpretation • Proposal for next steps	• Process Description, Design Basis, Philosophy • PFDs • Full process model • Material and Energy Balance • Economic analysis • Descriptions of key equipment, including critical design parameters • Technology risks and unknowns	• Process Description, Design Basis, Philosophy • PFDs • Process model • Material and Energy Balance • P&IDs • Equipment List • Economic analysis • Descriptions of key equipment, including critical design parameters • Operational Considerations • Technology risks and unknowns • Pilot Plant description, design, proposal	• Preliminary Equipment Design • Preliminary Layout • Preliminary Schedule • Analytical methods • Project Objectives • Site Selection • FlowSheets • Process model • Budget Estimate (30 - 40% accuracy) • Buy-in from all parties on option selection	• Process Description • Process models • Energy reduction opportunities • Yield improvement opportunities • Operability improvement • Prioritized list of improvement opportunities	• Purchase-ready Major Equipment Specifications • Project Execution Plan (procurement plan, project schedule) • Preliminary 3-D model • Electrical Equipment List • Line list • Complete P&IDs • Authorization Grade Estimate (10 - 20% accuracy) • Procurement Plan • Detailed Scope of Work

Mid-Atlantic Technology, Research & Innovation Center
MATRIC

5

Technology Transfer Process

FEL-1	FEL-2	FEL-3
Develop cost effective process design(s), including chemistry, separations, physical properties, and economics. Identify technology unknowns and demonstrate technology feasibility sufficient to justify moving to next phase.	Evaluate process design(s), economics, and operations to select the optimal process. Technology unknowns are identified and resolved.	Complete Basic Facility Scope Package to define process technology for Detailed Engineering, Procurement, and Construction.
Client Business project champion, client technical lead, MATRIC technical lead, MATRIC project leader	Client business leader, client project champion, client technical lead, MATRIC project leader, MATRIC technical lead	Client business leader, client project champion, client technical lead, MATRIC project leader, MATRIC technical lead

Proof of Concept	Laboratory Research	Conceptual Process Design(s)	Pilot Plant testing and validation	Basic Process Engineering Design	Design and Operations Optimization	Process Engineering Design

Mid-Atlantic Technology, Research & Innovation Center
MATRIC

6

Preliminary Definition of Technology that includes:	Definition of Technology that includes:
• Process Description, Design Basis, Philosophy	• Process Description, Design Basis, Philosophy
• Pilot Plant PFDs	• Preliminay Demonstration Plant PFDs
• Full process model	• Process model
• Pilot Plant Material and Energy Balance	• Demonstration Plant Material and Energy
• Economic analysis	• P&IDs
• Descriptions of key equipment, including critical design parameters	• Equipment List
• Technology risks and unknowns	• Economic analysis
• Inherent Safety Analysis, Pilot Plant HAZOP	• Descriptions of key equipment, including critical design parameters • Operational Considerations • Technology risks and unknowns

7

Preliminary Definition of Technology that includes:	Definition of Technology that includes:
• Process Description, Design Basis, Philosophy	• Process Description, Design Basis, Philosophy
• Pilot Plant PFDs	• Preliminay Demonstration Plant PFDs
• Full process model	• Process model
• Pilot Plant Material and Energy Balance	• Demonstration Plant Material and Energy
• Economic analysis	• P&IDs
• Descriptions of key equipment, including critical design parameters	• Equipment List
• Technology risks and unknowns	• Economic analysis
• Inherent Safety Analysis, Pilot Plant HAZOP	• Descriptions of key equipment, including critical design parameters • Operational Considerations • Technology risks and unknowns

8

What Are Technology Unknowns?

Scale Related Issues

Are issues identified a small scale going to be issues at commercial scale?

Are there issues that are only apparent at commercial scale?

Third Party Technologies

What are the assurances of performance?

What are the consequences of not meeting the target?

Inherent Technology Uncertainties

Is it a large step change from current technology?

What is the downside if performance is slightly off? Dramatically off?

Are their customer supply expectations that are critical? Volume? Quality?

Unknown Unknowns

MATRIC

Mitigating Technology Unknowns

- **Identify and Resolve Unknowns**
  - Lab studies
  - Pilot plants
  - Plant tests
- **Design Plant with a Wider Operating Window**
  - Design for higher flows, broader temperature windows, etc.
  - Design process to be more robust to process upsets
- **Prepare Engineering Packages as Contingency**
  - Packages prepared but not installed
    - High cost to install, low likelihood to be needed, downtime to install is acceptable
    - If cost is low, perhaps install the mitigation as a contingency
- **Install parallel systems**
  - Install conventional technology and innovative technology in parallel
    - Especially for the first of multiple applications
    - Also, if consequences of downtime are high

What about the Unknown Unknowns?

MATRIC

10

Design for Wider Operating Window

- **Description of Technology Opportunity**
  - Lower concentration of a reaction coproduct will lead to improved economics due to increased yield and reduced downtime
    - Product kinetics very sensitive to coproduct inhibition
    - Conventional removal technology treats a slip stream from the process
- **Potential Issues**
  - Removal technology is from an external licensor
    - Need to operate at conditions outside vendor's experience
    - Technology risks inherent in treatment technology not under project team's control
- **Consequences of Not Achieving Design**
  - Reduced production and raw material efficiency until resolved
  - Physical repairs will require months of downtime, lost sales, high modification costs
- **Mitigation**
  - Removal technology designed to treat entire stream, but piping configured to operate as a slip stream. Also makes process more robust.
  - Adds some excessive costs for larger piping, column to handle excess gas flow; higher operating costs from less than optimum operation

Mid-Atlantic Technology, Research & Innovation Center
MATRIC

11

Design for Wider Operating Window 2

- **Description of Technology Opportunity**
  - Designing a plant for a new, more selective catalyst
    - Will result in less heat removal needed
      - Can use larger reactor tubes
      - Will reduce size of reactor, saving steel and $millions
- **Potential Issues**
  - Underperforming catalyst
    - Too little area for heat transfer leading to reactor runaway
  - Catalyst scale-up not successful
- **Consequences of Not Achieving Design**
  - Plant shutdown and catalyst change (if underperforming batch) costing loss production, maintenance
  - Plant shutdown and replace reactor if catalyst scale-up issue costing $10's of million
- **Mitigation**
  - Design reactor tubes for conventional catalyst performance
    - Use conventional catalyst as back-up
    - Gives up some potential cost savings in reactor

Mid-Atlantic Technology, Research & Innovation Center
MATRIC

12

Contingency Planning

- **Description of Technology Opportunity**
  - A process reconfiguration has been identified which will result in 30% reduction in capital with operating and maintenance cost reductions as well
    - Process validated in pilot plant scale campaign
- **Potential Issues**
  - Product being recovered is a monomer
    - Polymer plugging of column a potential issue, and actually seen in pilot plant
      - Pilot Plant uses Oldershaw columns with <1mm gap between bottom of downcomer and next tray. Gap can plug
- **Consequences of Not Achieving Design**
  - Column plugging will lead to plant downtime, lost production, increased maintenance costs
- **Mitigation**
  - Use of dual flow trays to make the column less susceptible to plugging
  - Identify and scope more robust inhibitor distribution, in case it is necessary
  - Expertise for addressing new challenges on site before start-up

Mid-Atlantic Technology, Research & Innovation Center
MATRIC

13

Construct Parallel Processes

- **Description of Technology Opportunity**
  - A new catalytic reaction has been identified that significantly increases raw material yield
    - Conventional process is a homogeneous reaction without a catalyst
- **Potential Issues**
  - Catalyst has been demonstrated in the lab and piloted
    - Not possible to use actual raw materials
    - Final scale-up impacts unknown
- **Consequences of Not Achieving Design**
  - Plant shutdown and conversion to conventional design will be costly (both pounds and $)
- **Mitigation**
  - Since conventional reactor design is very inexpensive, install a homogeneous reactor parallel to the catalyzed reactor
    - Valve switching to allow ready reactor selection

Mid-Atlantic Technology, Research & Innovation Center
MATRIC

14

Wrap up – what challenges are

- **Successful implementations require management of the technical uncertainties**
  - Engineered mitigations
    - Parallel technologies
    - Excess capacity
  - Contingency Plans
    - Revisions designed, not implemented
    - Product supply plans
    - Plant instrumentation decisions
  - Staffing Plans
    - Availability of SMEs, including on start-up team
    - Support labs readily available through start-up

Mid-Atlantic Technology, Research & Innovation Center
MATRIC

15

Contact MATRIC

Jack Dever
Chief Technology Officer and Executive Vice President

Office: (304) 720-8617
Mobile: (504) 463-4372
jack.dever@matricinnovates.com

Mid-Atlantic Technology, Research & Innovation Center
MATRIC
Market-Driven Innovation ™

Mid-Atlantic Technology, Research & Innovation Center
MATRIC

16

503670 Digital Alchemy for Assembly Engineeringc

Abstract:

Back in the day, Stuart Churchill was a pioneer in the use of computers for modeling chemical and thermal processes. Since that time, the role of computing in chemical engineering has become indispensable. In this talk, we present new computational methods we call "digital alchemy" that allow for predictive modeling of thermodynamic assembly, and permit the design of building blocks optimized for assembly into target structures with desired properties. In this approach, standard thermodynamic ensembles are generalized to include alchemical variables and their thermodynamic conjugates, allowing for rapid, real-time computational investigation of large, complex design spaces.

503671 Improved Algebraic, Numerical, and Graphical Representations in Fluid Mechanics

Abstract:

The concept of equivalents is applied to fully developed flows through straight channels, and two applications are introduced: the equivalents that produce equal rates of flow through channels of different cross-section, and the diminishment that occurs between the regimes of laminar and turbulent flow. The validity of the eddy diffusivity and the invalidity of the mixing-length as measures of the shear stress due to turbulence have perhaps never before been demonstrated so clearly. The possible choices of primary variables and the consequences are also examined. The objective of these explorations is an improved understanding of the basic fluid mechanics by teachers, students, and industrial practitioners.

503672 The Scaling of Turbulence Near the Wall and the Churchill Turbulent Flux Correlation: Insights with Lagrangian Simulations

Abstract:

The evaluation of mean velocity and mean temperature in wall-bounded turbulent flows has been based on the "classical" theory of turbulence and the law of the wall. The viscous wall parameters, namely the friction velocity, the friction temperature, and the kinematic viscosity of the fluid have been used to scale not only the mean values of velocity, but also other mean turbulent quantities close to the wall, including the Reynolds stresses and the turbulent velocity and temperature fluctuations. Stuart Churchill, as he questioned the validity of the Reynolds analogy between momentum and heat or mass transfer in his works [1], he realized that extending the scaling of momentum to the scaling of scalar quantities in turbulence was not a sound approach. He proposed a scaling of turbulence that depends on the turbulent momentum flux and the turbulent heat flux. In his AIChE Institute Lecture (in Miami, 1998), Churchill presented a summary of his work on this topic, and closed form solutions for the mean velocity and temperature [2]. The turbulent Prandtl number, the most significant parameter for the modeling of turbulent transport not only in theory but also in modern CFD commercial software packages, was also interpreted physically based on these ideas [3]. We have used direct numerical simulation results and results from Lagrangian simulations [4] for the transport of passive scalars in turbulent channel flows to explore Churchill's ideas on turbulent flow scaling and to develop an updated theory for the scaling of heat transport in turbulent flows [5-7]. His passion to remove empiricisms and approximations from the development of predictive correlations in turbulent transport has been an inspiration. In this talk, we will revisit the Churchill scaling in view of our simulation results, and will discuss the counterintuitive finding that wall turbulence can be used to separate rather than mix substances under special circumstances [8].

1. Churchill, S. W., 1996, "A Critique of Predictive and Correlative Models for Turbulent Flow and Convection," Ind. Eng. Chem. Res., 35, pp. 3122–3140.
2. Churchill, S.W., 2000, "Progress in the Thermal Sciences: AICHE Institute Lecture," AIChE Journal, 46(9), pp. 1704-1722.
3. Churchill, S. W., 2002, "A reinterpretation of the Turbulent Prandtl Number," Ind. Eng. Chem. Res., 41, pp. 6393–6401.
4. Papavassiliou, D.V, 2002, "Turbulent transport from continuous sources at the wall of a channel," Int. J. Heat Mass Transfer, 45(17), pp. 3571-3583.
5. Le, P.M., and Papavassiliou, D.V., 2008, "On the scaling of heat transfer using thermal flux gradients for fully developed turbulent channel and Couette flows," Int. Commun. Heat Mass Transfer, 35(4), pp. 404-412.
6. Srinivasan, C., and Papavassiliou, D. V., 2011, "Prediction of Turbulent Prandtl Number in Wall Flows With Lagrangian Simulations," Ind. Eng. Chem. Res., 50(15), pp. 8881–8891.
7. Srinivasan, C., and Papavassiliou, D.V., 2013, "Heat transfer scaling for wall bounded turbulent flows," Applied Mechanics Review 65(3), Art. 031002.
8. Nguyen, Q., Srinivasan, C., and Papavassiliou, D.V., 2015, "Flow induced separation in wall turbulence", Phys Rev E, 91, Art. 033019.

503674 Flow Boiling Using a Piranha Pin Fin Heat Sink

Abstract:

Flow Boiling using a Piranha Pin Fin Heat Sink

Corey Woodcock, Xiangfei Yu, Yoav Peles[1] and Joel Plawsky

Rensselaer Polytechnic Institute; [1]University of Central Florida

The *Piranha Pin Fin* (*PPF*) microdevice is a heat exchanger engineered to dissipate extreme thermal loads using either dielectric fluids such as HFE7000 or water. Fabrication of the device is CMOS compatible and the first generations of the device have been fashioned from silicon substrates. The PPF operates on the principle that heat transfer would be more efficient if heated fluids could be removed nearly as fast as they are created. This process would keep the driving force as high as possible. Thus the heat exchanger is built as two layers; a primary substrate that dissipates the bulk of the heat and shunts the heated fluids to a booster section that continues to heat the fluid and extends the total surface area of the device. Three generations of devices were fabricated. The first two generations were single-layer systems designed to help understand the operation of the primary layer and optimize the shape and location of the pin fins. The third generation was the first, two-layer system to be fabricated and tested.

Figure 1 Representative PPF array

Correlations for the heat transfer coefficient, h_{tp}, based on the Boiling Number and the Weber Number in the PPF design was developed and validated with Gen II. The discrepancy of heat transfer coefficient between predicted and measured values was within 20%. The ratio of two-phase to single-phase pressure drop was correlated with the Boiling Number (Bo) and the Jakob Number (Ja). The discrepancy of the two-phase multiplier between prediction and experimental measurements were within 12%.

The third generation *PPF* heat sinks were the first series to provide both a primary substrate populated with *PPF*s, as well as a booster substrate populated with solid micro pin fins. The substrates are micro-aligned and fusion-bonded to provide a full three dimensional (3D) architecture with bulk silicon properties. Experimental results from the Gen III *PPF* microdevices provide a proof of concept of the true *PPF* heat sink concept with fluid venting into multiple substrates. At low flowrates (G = 670 [kg/s/m^2]) the Gen III *PPF* heat sink has been shown to dissipate heat loads approaching 1.1 [kW/cm^2] at an exit quality of 95%, while at higher flowrates (G = 3,300 [kg/s/m^2]) the thermal limit of the heat sink remains unknown due to exceeding the available power supply at >1.4 [kW/cm^2].

503673 Transport Problems in the Spirit of Stuart Churchill for Teaching and Research at the University of Michigan

Abstract:

Stuart Churchillâ€™s solutions to important heat transport problems have greatly influence both pedagogy and research. The correlation of Churchill and Chu for the natural convection heat transfer to or from a vertical plate, over an astonishing range of Rayleigh numbers is based partly on a similarity solution of a type common in heat, mass, and momentum transfer for problems in chemical engineering. One such problem is asphaltene or wax transport and deposition onto the wall of an oil pipeline, which can be solved using similar methods, as will be discussed. In addition, such problems, and the wealth of their applications make for excellent motivations and examples in the teaching of graduate or undergraduate transport, as will be discussed in the talk.

509054 Solve this! Fundamental Approach to Problem Solving in Industrial Processes I (Invited Talks)

Abstract:

Solving complex plant problems is an everyday exercise that affects all projects and budgets, and it is very rarely introduced to students at any level. In fact, by the time young engineers are few years into working in industry, most have lost the engineering skills that they developed in school. We hope to encourage future engineers to retain those skills and learn the importance of plant problem-solving.

497393 From Atom Groups to Molecules and Mixtures/Formulations: A Comprehensive Design Methodology with Generalized Disjunctive Programming

Abstract:

In the chemical industry, the design of mixtures and blends that meet a specified set of physical and chemical properties is an important and challenging activity across a wide range of applications. Several chemical examples with different functions and qualities, such as agrochemicals (pesticides, insecticides, etc.), functional chemicals (solvents, refrigerants, lubricants, etc.) and household products (cosmetics & personal care products, pharmaceuticals & drugs, etc.), come as formulations/mixtures and are used in most aspects of human life. Computer-Aided Molecular (CAMD) and Mixture/blend (CAM^{b}D) Design[1,2] is a promising approach for efficient and reliable design of promising molecules, mixtures or blends that meet predefined target properties and optimise a given performance measure. When formulating a mixture problem via optimisation, the main purpose is to identify innovative solutions, which can be a new molecular structure, an existing molecule for a new property function, or a new mixture of existing/new molecules. Mixture design, however, can be very challenging because it requires finding the optimal number, identities and compositions of mixture components, and using nonlinear property models, which results in complex nonlinear problems with a large combinatorial space. Thus, a reduced version of the general CAM^{b}D problem is usually posed: the number of mixture ingredients is fixed *a priori* and the identity of a compound (or of all compounds) that can participate in a mixture is selected from a restricted list of candidate compounds, leading to suboptimal solutions.

In view of these challenges, our work focuses on developing a comprehensive methodology within the computer-aided mixture/blend design framework by combining mathematical modelling, optimisation and chemical engineering insights to formulate the general mixture problem. Within this approach, the optimal number of components in a mixture, their identities and compositions are determined simultaneously[3,4] and the desired molecules are designed from a large set of atom groups (UNIFAC groups). A logic-based methodology, Generalised Disjunctive Programming[5,6] (GDP), is used to express the general mixture problem within a mathematical framework and formulate the discrete choices inherent in the problem (i.e., how many components are designed, which specific molecules should be used - what atom groups are required). In order to exploit existing MINLP algorithms, Big-M approach[7] is employed to transform the disjunctive constraints into mixed-integer form.

The proposed framework has been applied successfully to two case studies where the design of solvent mixtures for separation processes is presented. The first case study involves the design of optimal solvent and antisolvent mixtures for cooling and drowning out crystallization, respectively. In the second case study, optimal solvent mixtures are determined to separate acetic acid from water in a single stage liquid extraction process. Finally, integer cuts are introduced to the general mixture formulations and a list of optimal solutions (i.e., list of mixtures with different number, identity and compositions of ingredients) is obtained for each problem. Significant benefits can accrue by employing the general framework (number of mixture constituents not fixed a priori and molecules designed from functional groups) in mixture and product design: avoid evaluating explicitly every choice of the number of components, which can be computationally or experimentally costly and time-consuming, especially as the number of desirable ingredients increases; consider larger design spaces where many molecules and mixtures are designed and not selected from a limited set of choices.

1. Achenie LEK, Gani R, Venkatasubramanian V. *Computer aided molecular design: theory and practice.* Amsterdam, The Netherlands: Elsevier. 2003.
2. Gani R. Chemical product design: challenges and opportunities. *Computers & Chemical Engineering.* 2004;28:2441–2457.
3. Jonuzaj S, Akula PT, Kleniati PM, Adjiman CS. The formulation of optimal mixtures with Generalized Disjunctive Programming: A solvent design case study. *AIChE Journal.* 2016;62:1616–1633.
4. Jonuzaj S, Adjiman CS. Designing optimal mixtures using Generalized Disjunctive Programming: Hull relaxations. *Chemical Engineering Science.* 2017;159:106–130.
5. Balas E. Disjunctive Programming. In: *Discrete Optimization II*, edited by Hammer PLK, Johnson EL, Korte BH, vol. 6 of *Annals of Discrete Mathematics*, chap. 6, pp. 3–51. Amsterdam, The Netherlands: Elsevier. 1979.
6. Raman R, Grossmann IE. Modelling and computational techniques for logic based integer programming. *Computers & Chemical Engineering.* 1994;18:563–578.
7. Nemhauser GL, Wolsey LA. *Integer and combinatorial optimization*, vol. 41. New York: Wiley & Sons, Inc. 1999.

497375 A Novel Computer-Aided Model-Based Tool for Chemical Product Design

Abstract:

In chemical product design, the application of model-based methodologies can help to design/improve products so as to reach the market faster by reducing costly and time-consuming experiments at the early stages of design [1]. That is, experiments are only performed during the last stage as a verification step. Since product candidates involve thousands of mixtures/blends of chemicals that need to be evaluated, a huge amount of data on physico-chemical properties of chemicals and their mixtures and/or models that can reliably predict them are needed for tailor-made design of products. This is a challenging task requiring data acquisition, data testing, model development, model-based design method development, etc, that needs to be integrated within a computer-aided framework so that chemicals based products can be designed, analyzed and verified in a fast, efficient and systematic manner [2].

In this work, the development of a systematic model-based framework for product design and evaluation, implemented in a new product design simulator called ProCAPD is presented. The chemical product simulator works in the same way as a chemical process simulator, but has additional features. Like a process simulator, the chemical product simulator is able to perform calculations on the specified product. Instead of the process flowsheet, the product molecular structure (in the case of a single molecule) or a mixture of molecules is given as input. Analogous to the process conditions, the product application conditions are specified and similar to the mass and energy balance calculations in the process simulator, the properties-functions of the product are calculated. Like the contents of the process simulator, the product simulator needs a database of chemicals and properties, a library of models, numerical routines to solve mathematical problems as well as various calculation options. Also, like the process simulator, the product simulator comes with a user-interface to describe the problems and to obtain the simulation results.

In order to make the chemical product simulator versatile with a wide range of applications, it includes a suite of a built-in knowledge base that has a suit of databases containing properties of different classes of chemicals (lipids, solvents, aroma, polymers, cosmetics, etc.) and blended products (gasoline, jet-fuels, lubricants, hair spray, etc.). It can calculate 55 pure component properties (such as critical properties, acentric factor and solubility parameter) and 10 functional properties (such as vapor pressure, liquid viscosity, and surface tension) and includes databases covering more than 24,000 compounds. Also, it has models for phase equilibria predictions such as UNIQUAC, UNIFAC, NRTL and PC-SAFT. It has special model parameters and data for lipids and ionic liquids. Transport related properties and new interaction parameters for a UNIFAC model for lipids systems have also been incorporated. These features together allow the design of a very wide range chemical-based products. The large amount of data, models and calculation-algorithms are managed through a specially developed ontology. All these capabilities are based on the prototype tool developed by Kalakul et al. (2017) [3].

The presentation will highlight the software architecture, the implemented computer aided methods-tools and the scope-significance will be illustrated through new chemical product design-evaluation applications.

References

[1] M. Hill, 2009. Chemical Product Engineering â€“ The third paradigm, Computer & Chemical Engineering, 33(5), 947-953.

[2] R. Gani, 2004, Chemical Product Design: Challenges & Opportunities, Computers and Chemical Engineering, 28, 2441-2457.

[3] S. Kalakul, S. Cignitti, L. Zhang, R. Gani, 2017, Chapter 3 â€“ VPPD-Lab: The Chemical Product Simulator, Tools For Chemical Product Design, Computer Aided Chemical Engineering, 39, 61-94

501375 Antibiotics Molecular Design Using Artificial Bee Colony Optimization

Abstract:

Antibiotics Molecular Design Using Artificial Bee Colony Optimization

Authors: Shweta Mapari, Matthew Hartenstein, Rex Gaumer, Kyle Camarda

In a recently published WHO report, a list of bacteria which pose the greatest threat to human health was given. The ESKAPE list also defines pathogens which have become resistant to multiple drugs and for which there exists a critical need for alternate therapies. Acinetobacter *baumannii,* which is on both of these lists, is a pathogen causing infections such as pneumonia and bacteremia. The goal of this research is to apply Computer Aided Molecular Design (CAMD) techniques to develop a novel antibiotic agent to treat this bacterial infection.

In order to treat A. *baumannii*, two molecular design problems were formulated, one for a beta lactam antibiotic and the other for a fluoroquinolone compound. CAMD is used to find analogous structures of small molecules which have reduced values of the Minimum Inhibitory Concentration (MIC) and toxicity for beta lactam and flouroquinolone drugs. We used piperacillin, amoxicillin, and ticarcillin as starting points for the beta lactam drugs and paufloxacin as a starting point for the fluoroquinolone drugs.

This work applies topological molecular descriptors to develop Quantitative Structure Activity Relationships (QSAR) which are employed to predict MIC values and toxicity. These expressions are combined with structural feasibility constraints to formulate an optimization problem. Because of the large number of potential chemical structures and the uncertainty in the structure-property correlations, stochastic algorithms are preferred to solve the resulting MINLP. One stochastic method which has shown promise to solve these problems is the Artificial Bee Colony algorithm, which relies on principles of swarm intelligence to find near-optimal solutions efficiently. The Artificial Bee Colony algorithm described in this paper is used to derive solutions which serve as lead compounds for a narrowed search for novel antibiotics, which would employ molecular simulation and experimental validation.

500584 An Integrated Methodology for Chemical Product Design: Application to Hair-and Skin-Care Emulsions

Abstract:

Due to the large number of ingredients commercially available for the formulation of consumer products, and because their multi-functionality, emulsified cosmetic product design is a challenging task. The traditional design approach often relies on costly and time consuming trial-and-error experimental procedures. In recent years, systematic approaches to reduce time and resources using computer-aided mixture/blend design (CAM^{b}D) have been proposed [1-2]. Based on predictive models for product properties and mixed-integer optimization, these design methodologies are able to explore a large domain of possibilities. They are however of limited applicability to few formulated products due to the lack of reliable models for properties prediction or for synergic effects.

In particular industrial sectors such as the cosmetics one, there is a considerable amount of heuristic knowledge, namely regarding qualitative function of ingredients, their incompatibilities and positive synergies, as well as their impact on sensorial attributes, which is a critical aspect for product acceptance [3]. This heuristic knowledge has been recently incorporated into a systematic CAM^{b}D methodology [4] using mixed-integer optimization with logical constraints [5] and in this work is further developed for cosmetic emulsions. The method considers a list of available ingredients classified according to their main function (e.g., emollients, emulsifiers, rheology modifiers). Then, heuristics regarding ingredientsâ€™ selection, their concentrations in final product, and the manufacturing procedures are formulated as logical conditions, and further translated into algebraic constraints. These constraints, together with other known restrictions (technical, toxicological and/or legal, etc.), define a reduced design space. The search in this space is guided by property models relating product composition to key physicochemical properties or sensorial attributes. The multi-objective function is composed of a weighted sum of Taguchi functions with a target value for each cost/performance requirement (e.g. softness, greasiness and thickness of the product, etc.), which could be adapted by the designer. Through integer cuts, the entire set of feasible solutions within the reduced design space or a subset of interest may be generated.

To illustrate the proposed methodology, two case studies of a rinse-off conditioner and a body lotion are presented. From an initial list of 24 emollients, 4 rheology modifiers and 8 emulsifiers, a set of alternative formulations and respective manufacturing procedures is listed. The first 5% of these resu-lting alternative formulations were produced and then analyzed using instrumental and sensorial methods. The predicted values of the performance were similar for most of the formulations, confirming the potential of this methodology to model emulsified products.

REFERENCES

[1] L. E. K. Achenie, R. Gani, V. Venkatasubramanian, Eds., *Computer Aided Molecular Design: Theory and Practice*. Amsterdam: Elsevier, 2002.

[2] M. Mattei, G. M. Kontogeorgis, R. Gani, «A Systematic Methodology for Design of Emulsion Based Chemical Products», In: Karimi IA, Srinivasan R, eds., *Computer-Aided Chemical Engineering*, vol 31. Amsterdam: Elsevier, 2012, pp. 220-224.

[3] A. M. Pensé-Lhéritier, «Recent developments in the sensorial assessment of cosmetic products: A review», *Int. J. Cosmet. Sci.*, 2015, vol. 37, pp. 465-473.

[4] J. Arrieta-Escobar, F. Bernardo, A. Orjuela, M. Camargo, L. Morel, « An Integrated Methodology for Emulsified Cosmetic Product Formulation Using Integer Programming with Logical Constraints», (Provisionally accepted paper for ESCAPE 27)

[5] H. P. Williams, *Model building in mathematical programming*, 5th ed. John Wiley & Sons, 2013.

491188 Product Design: A Systematic Pricing Framework for Maximizing Company Profits

Abstract:

The importance of chemical product design has been well-recognized in the chemical engineering community. Many articles have been published to consider various aspects in product design, including material selection, product conceptualization, and process design to meet the desired product attributes. However, pricing strategy and pricing decisions receive little attention in the chemical engineering community. This is a serious omission as product pricing is important for a company to set the optimal price for getting a reasonable return and at the same time has a good market penetration.

A systematic pricing framework has been developed for pricing a chemical product. First, a pricing strategy is selected based on product characteristics, company strategy, and market situation. Market information and company data are collected next. The collected data are used to predict the parameters in a pricing model. For example, advertising data are used to estimate the product awareness in the market, whereas the consumer preference data between our product and the competing product is used to estimate whether the consumer will purchase our product or the competing product. With all these information, a relation between product price and product demand can be obtained from a pricing model. Finally, the optimal product price is obtained by maximizing the companyâ€™s profit. This framework is illustrated with an energy drinks example. With this framework, product design can be optimized not only from a technical perspective, but also from a consumer perspective and a business perspective.

"We Don't Expect It to Be a Problem": Best Practices in Pilot Plant Start Up

Richard Palluzi

Richard P Palluzi LLC

Laboratory, pilot plant and research safety, design, training and consulting

72 Summit Drive, Basking Ridge, NJ 07920
rpalluzi@verizon.net 908-285-3782

1

Introduction

- Pilot plant start ups are very different from plant start ups
  - Sadly, rarely recognized
  - Nor are the differences understood and evaluated in advance to address problems in approach
- Major differences
  - Scale of the unit is enormously smaller
    - \+ Assumed proportionally easier
      - o While less, it's not proportionally less whether based on size, scale, or capital cost
    - \+ Produces false sense of security
    - \+ Discourages careful advance planning
  - Resources committed are proportionately much less
    - \+ Effort required, while less overall, is significantly higher than scale suggests
    - \+ Range of expertise assigned to the start up team is much more limited
    - \+ Start up effectiveness becomes inordinately dependent on the skill of the few personnel involved
    - \+ Competing time draws destroy efficiency and effectiveness
  - Operation is often novel or different
    - \+ Limited operating experience makes analysis difficult
    - \+ Increases troubleshooting difficulty exponentially
    - \+ Makes even identifying problems significantly harder
- Realistic, advance start up planning is rarely performed

"We don't expect it to be a problem"

Words that frequently come back to haunt you ...

2

Let's Define The Elements of A (Realistic) Successful Pilot Plant Start Up

1. Safety – no injuries or incidents
2. Quality – the unit's capabilities are fully understood and it performs close enough to the design basis to run the proposed program
3. Schedule – completed as fast as possible given the resources committed
4. Cost – as little as possible given the schedule constraints

- Please note:
  - Despite your best efforts the pilot plant may not do everything you hoped it would do or at least not as well as you wanted
    - But it does it well enough to start the experiments or tests
  - There may be unpleasant surprises lying in wait for you down the road but you feel confident that you should be able to get usable data for at least the first phase(s)
    - This tacitly assumes that you have not changed the programs upon which the pilot plant's design was based
  - The chemistry or process may not do what you expect
  - It will not be fast
  - It will not be cheap

3

The Key Elements to A Successful Pilot Plant Start Up

- Careful advance planning
- Assembling the right team
- Developing a detailed start up plan
- Making a realistic schedule and cost
- Not giving up

4

Careful Advance Planning

- Review the proposed design several times for potential start up problems
  - Before appropriation
    - So you have enough money to do the right thing
  - Before construction
    - So you don't have to do it twice
    - So you are not constrained by choices very difficult to change
  - Before construction completion
    - So no construction issues go unrecognized until it's too late
  - Independent reviews or cold eye are most valuable
    - Because they ask the questions we are afraid to raise
    - Because they don't share the same prejudices and preconceived ideas
  - Need to question everything
    - How do I turn it on?
    - How do I calibrate it?
    - How do I confirm it's working?
    - Do I need anything to do these tasks?
      - Supplies
      - Manuals
      - Training
      - Consumables
      - Etc.
- Resolve them all before construction starts (or at least finishes)

> *"We had identified that as a potential disaster during the design but hoped we were wrong as changing the design to prevent it would have been expensive and time consuming"*

5

Typical Problem Areas – A Very Brief Sample

- Is appropriate valving provided for calibration and maintenance?
- Can all parts of the system be drained, flushed and/or purged easily?
- Can the system be leak tested easily?
- Are there any components with an unacceptably high leak or failure probability?
- Is every component accessible?
- Are sufficient indicators or sensing devices provided for validation and verification?
- What fallback positions are available for problem areas?
- Are adequate spare parts on hand or on order?
- How will each system reach operating conditions?
- How can every instrument be calibrated?
- Is any special calibration equipment or services required?
- Who knows how to start up any novel equipment?
- Is adequate equipment documentation available or on order?
- How will off spec product, flushing media, etc. be handled?
- Who has determined all the instrument spans, are they available, and are they correct?
- Are any special provisions required for testing?

6

Careful Advance Planning (cont'd)

- Review past start ups
  - Identify the major problem areas
  - Evaluate what practices worked and what practices didn't
  - Discuss potential better practices
  - Develop honest lessons learned
  - Look for the actual issues not the officially blessed ones
    + Often requires you to talk to the people directly involved off the record
    + Recognize that it may not be in your power to prevent the same mistakes again
    + But at least you can not be surprised
- Common lessons learned
  - Inadequate advance planning
  - Inadequate resources
    + Quantity and/or time commitment
  - Inexperienced personnel
  - Lack of adequate craft or specialist support
  - No central direction
  - Unrealistic schedule and/or budget
  - Major shifts in the research program during design and construction
  - Lack of proper testing and calibration equipment
  - Delayed realization that no knows how to set up or calibrate something
  - Lack of any realistic commissioning plan
  - Purchase of novel equipment with no advance testing
  - Lack of viable fallback positions for critical areas
  - Inadequate start up QA/QC resulting in significant rework and blind paths
  - Lack of realistic procedures for a task resulting in rework
  - Failure to understand how a system or equipment is supposed to work
  - Lack of adequate troubleshooting information or data

7

Careful Advance Planning (cont'd)

- Start planning the start up early
  - Arrange design reviews before the design is finalized
    + The reviews take time
    + Discussion is often required to determine how to best address the concern
  - Develop the start up plan well before construction ends
    + Make sure everyone reviews it and understands it
    + Make sure everyone buys into it and knows their part in it
  - Review the construction plan to identify opportunities for concurrent or overlapping work to save time
    + Be realistic as many plans often put everyone in each other's way
- Develop a detailed start up plan

8

Start Up Plan

- Detailed list of all tasks you can identify need to be completed to finish the start up
  - Should be specific to the unit not generic
  - Should be as detailed as possible
  - Should not assume anyone knows what to do unless spelled out
  - Will take time and effort to develop
  - And time to revise to get consensus
- The plan should
  - Indicate who will be responsible for performing the task
  - How long the task is estimated to take
  - This allows realistic resource evaluation, scheduling, and tracking
- The plan allows you to
  - Identify many problem areas in advance
  - Determine the required resources or the effect of limited resources
  - Optimize the work both in sequencing or scheduling
  - Make sure everyone knows what they are expected to do and by when
- Helps avoid unrealistic expectations

"I didn't know I needed to do it that way"

"Since construction will not be finished until a week before we promised to start the testing, we have planned for a 1 week start up"

9

Some Examples of Individual Tasks From a Start Up Plan

Sequence Number	Task Description	Responsibility	Estimated Duration (hrs)	Comments
1.9	Make up valve tags per Company Standard XYZ	Operator	8	Tagging kits available from Operations. Complete at least 1 week before project construction completion
18.2	Fill compressor with oxygen compatible fill	Project Eng	0.5	Check with Mfr who promised answer by Nov 1st. Confirm fill meets ASTM G63 standards. Clearly tag unit when complete. Make sure at least one extra oil change is on hand.
36.1	Perform loop testing on T107 i. Disconnect at TC and connect Type K simulator set at 300 °C ii. Confirm reading on computer screen iii. Set T107C to manual and vary between 0-100% output iv. Confirm CV 107 moves from full closed to full open	Instrument Tech	1.0	Make sure calibrator has been certified within last 3 months.

10

Common Major Areas Of A Start Plan

- Tracing
- Tagging
- Flushing
- Leak testing
- Completing insulation
- Leak testing heated/cooled components at temperature
- Instrument calibration
- Loop testing
- Validating instrument reading
- Validating control loop operation
- Safety system check out
- Confirming rotating equipment operation
- Conduit sealing
- Confirming ventilation
- Tuning
- Commissioning in sections
- Dummy or preliminary run
- Conducting final safety and operational views
- Updating all drawings

11

Careful Advance Planning (cont'd)

- Use the start up plan to estimate the quantity and type of resources the plan requires
- Develop a realistic schedule based on
  - The actual resources you can make available
  - Include holidays, business trips, and any scheduled commitments
  - All the identified work in the plan
- This is probably going to be harder than you realize as how long it takes to do anything is inordinately dependent on the doer's skill level and experience
  - Limited team size means averaging effect (as on a large project) is not possible
  - Often unknown at the scheduling stage
  - Leaving you to either assume an "average" time or guess the skill and experience
  - Very difficult to revise when the actual personnel are assigned
  - May want some advance discussion with your management
- Always add a very healthy contingency for the unexpected
  - Which is a significant amount since:
    - + It often takes significant work to find the cause of the problem
    - + Some problems requires time to develop and implement an effective solution

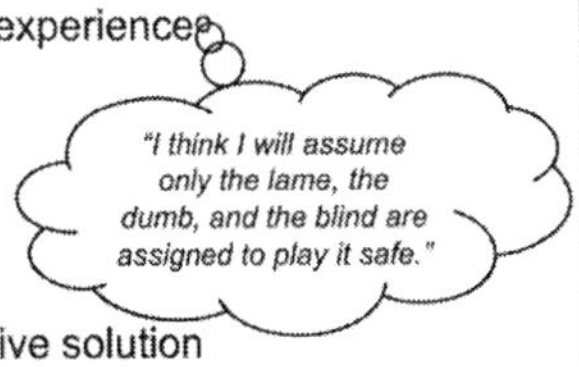

Pilot Plant Start Up Schedule Contingency

Team Experience Level	Unit Complexity	Process Familiarity: Well known	Process Familiarity: New and novel
Low	Low	40%	120%
	Medium	50%	150%
	High	75%	225%
Medium	Low	30%	60%
	Medium	40%	80%
	High	60%	120%
High	Low	20%	30%
	Medium	30%	45%
	High	50%	75%

(Time in detailed start up plan)(%above) ≈ time to add to start up schedule for unexpected issues

12

Start Up Schedule

- Once the start up plan is complete it can be used to generate a realistic start up schedule based on
  - The actual resources you can make available
  - All the identified work in the plan
- And, since it's realistic, you can then use it to track the progress of the effort

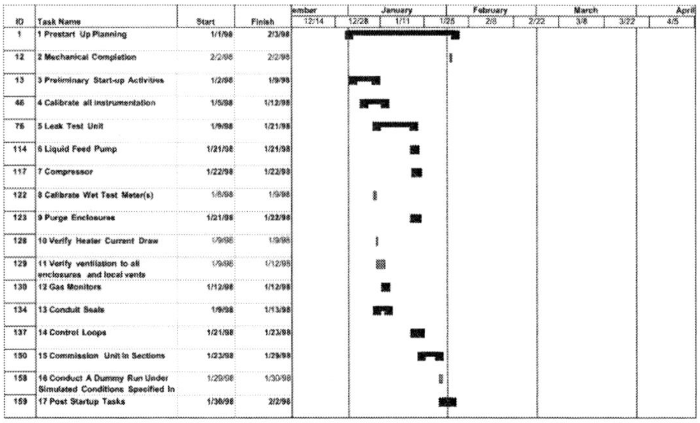

Please note these are rolled up task headings (so there are a lot more details than shown)

13

The Start Up Team

- Put <u>one</u> person in charge of the start up
  - You need a single driving force to get to completion
- Ensure the start up team has or at least has immediate access to adequate expertise
  - ✓ Instrumentation
  - ✓ Electrical
  - ✓ Piping
  - ✓ Process
  - ✓ Computer
  - ✓ Control
  - ✓ Mechanical hardware
  - ✓ Analytical
- Try to arrange for the team to be full time
  - Part time participation slows down progress disproportionately

14

Calibration

- Ensure all required test and calibration equipment is on hand before starting
  - Make sure it works
  - Make sure someone knows how to use it
  - Make sure it's adequate for the task
  - Make sure it is available
- Make sure the equipment is calibrated adequately
  - Is it accurate enough?
  - How was this accuracy determined?
  - How was the accuracy maintained?
  - How are you going to confirm or validate the calibration?
- Personnel
  - Are the personnel adequately trained?
  - Are they the ones assigned to the start up?
  - If extra training is required, has it been arranged in advance?
  - How knowledgeable is the person determining competence?
- Are any special calibration services required?
  - Have the vendors been checked and validated?
  - What means of double checking or confirming their work provided?

15

Miscellaneous Things Often Overlooked

- Settling on the initial instrument spans
- Getting the as built drawings on day one

  If the project team doesn't starts until the day before completion, prepare to be disappointed
- Tracing/confirming all the lines
  - And addressing unpleasant surprises
- Making sure any key materials are on hand before starting
  - Flushing media
  - Leak testing media
  - Feedstocks
  - Catalysts
  - Cal gases
- Scheduling final safety and operational reviews

  "If we had a realistic start up schedule and tracked it that would have been less of a disaster"
  - In advance so that you don't get delayed
  - But not so soon you are not finished
- Writing preliminary operating Instructions
- Creating run sheets or reports
- Settling on the preliminary experimental plan

  Think herding cats
- How you are going to go from a cold start to operation
- How you are going to go from operation to shut down

16

What Does It Cost

- Start up means different things to different people
- This accounts for variations in perceptions, schedules and costs
- Mechanical start up
  - Starts with construction completion
  - Lasts until all equipment energizes, moves or indicates
- Commissioning
  - Lasts until all major systems perform to specification
- Validation
  - Lasts until research personnel are convinced data generated is meaningful

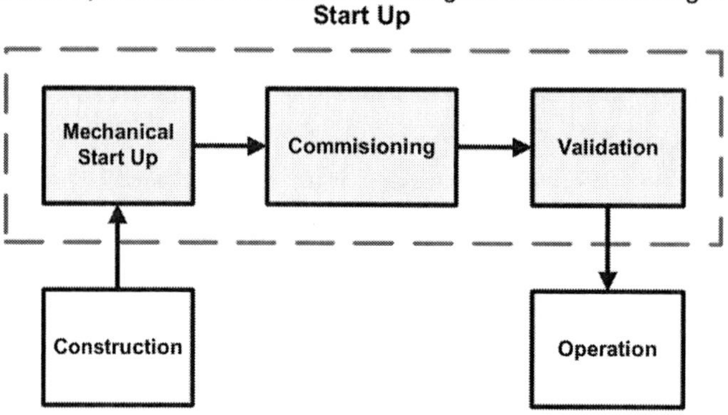

17

What Does a Typical Start Up Cost?

Unit Complexity	Start Up Costs to Construction Costs
Low	6% to 15%
Medium	10% to 30%
High	25% to 50%

Caveats

1. Covers mechanical start up and commissioning only.
2. Highly variable with ± 50% ranges common.
3. Assumes all costs are charged to the construction and to the start up (little to no to overheads).
4. Assumes good pilot plant experience.
5. Assumes well designed unit.
6. Assumes detailed start up plan made in advance.

Larger units usually cost less; smaller units may cost more.

18

Summary

- Pilot plant start ups are never easy
  - In fact they are much harder then most people – even experienced research people – realize
- With good advance planning, proper team selection, and a detailed plan it can be much more efficient, cost effective, and less painful
  - But never cheap, fast, nor easy

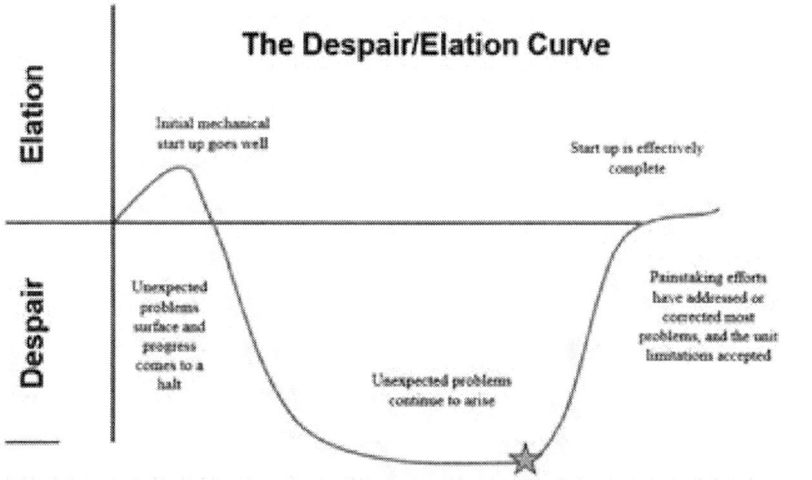

☆ *"When it appears that the unit can never be started up you are probably approaching completion." R P Palluzi*

19

499977 Optimal Steady-State and Dynamic Design of Experiments in Pilot Plants for CO_2 Capture

Abstract:

Testing and evaluation of novel CO_2 capture technologies in pilot plants are necessary for reducing the risk and uncertainty of applying these technologies at the commercial scale. However, pilot plant tests require significant time and resources, especially for the comparatively large power plant applications. Therefore, pilot plant test runs should be maximally informative. The typical approach to steady-state design of experiments (DOE) is to use a space-filling approach that considers only the input design space. Even when the output space is considered, a systematic approach is often not followed. Furthermore, typical experiments fail to exploit the knowledge gained as new data are collected. Thus, a sequential DOE with the potential for recourse during the test campaign can be advantageous for accelerating learning from the test runs. To develop a DOE that considers both the output space and enables a sequential DOE requires a mathematical model. As one of the typical test run objectives is to gather knowledge that may be missing or poorly represented in an existing model, it is expected that the existing model is highly uncertain, and new data collected from the test runs are expected to contain measurement uncertainty. Therefore, the sequential DOE should be developed with due regard of the uncertainty in the model as well as the measurements. With this motivation, a sequential Bayesian DOE is developed for steady-state test runs. Since our objective is to minimize the worst prediction variance of the existing model in the design space, the DOE is developed to be G-optimal.

In addition to steady-state DOE, a systematic approach to dynamic DOE for CO_2 capture pilot plants is also developed. However, dynamic tests can not only provide significantly more information than the steady-state test runs in a shorter period of time, but they can be instrumental in observing model parameters that cannot be observed through the steady-state test runs alone. Dynamic pilot plant test campaigns, if at all undertaken, typically consist of observations made during variable step changes. However, a dynamic DOE must ensure persistence of excitation so that all frequencies can be observed in the transient data. A D-optimal hybrid dynamic DOE is developed by designing a maximum length pseudo-random binary sequence coupled with a Schroeder-phased input.

The steady-state and dynamic DOEs are implemented in the Pilot Solvent Test Unit (PSTU) at the National Carbon Capture Center (NCCC) in Wilsonville, Alabama. Due to practical limitations of analyzing liquid samples and collecting experimental data, a batch design approach is developed. For tractability of the computational approach, surrogate response surface models are developed for use within the Bayesian framework. This presentation will include a critical analysis of the knowledge gathered during these test runs in comparison to the conventional DOE undertaken by the team in the recent past.

492154 Demonstration of a Commercial Scale Circulating Fluidized Bed Gasifier for Industrial Fuel Gas

Abstract:

Coal is the main energy resource of China and will continue to be predominant in the next few decades. Coal gasification has attracted worldwide interest due to its high efficiency and low pollution emissions. With the steadily rising demand in clean fuel gas, circulating fluidized bed (CFB) gasification has become one of the most popular gasification technologies in Chinese market. Some advantages of CFB gasification are that it can use low-cost pulverized coal, has satisfied adaptability of a wide range of solid fuels, can use poor quality coals and emits lower amounts of pollutants.

In 2015, Institute of Engineering Thermophysics (IET) had designed and built up 5 sites of CFB gasification system in Guangxi province. The designed fuel gas production capacity is about 40000 Nm^3/h for one site. The produced clean fuel gas is used to provide heat for alumina calcination. Novel key technologies were developed and carried out in this system, including counterflow air cap, air preheating technology and direct slag disposal technology.

In 2016, different types of low rank coals were used as feedstocks to study their gasification performances and operation stability of the CFB gasifiers. More than 320 days of long-term and stable operation has achieved, and significant experimental results are obtained to improve gasifier design and operation. For example, Shenhua lignite coal, and blended coal of Shenhua with Baikuang coal were investigated during industrial trials. The system temperature and pressure are stable, and no slagging phenomena occur. When shenhua coal is used as feedstock, the cold gas efficiency, the carbon conversion and the gas heat value of Shenhua coal gasification are 71.96%, 86.90% and 5.59 MJ/Nm^3. With the same air flow rate (about 18700 Nm^3/h), the cold gas efficiency, the carbon conversion and the gas heat value of blended coal gasification are 63.91%, 81.24% and 5.35 MJ/Nm^3. Mainly due to the poor gasification reactivity of Baikuang coal, the gasification performances of blended coal are lower than shenhua coal, but the produced fuel gas could satisfy the production of the alumina calcination. When Baikuang coal is added to Shenhua coal, the ash fusibility characteristics of Shenhua coal could be improved, which is helpful to achieve the long-term operation of industrial gasifier. The results indicate that CFB gasification is an efficient and clean coal conversion technology with low cost and high social recognition.

497797 Collaborative Bioprocess Development and Piloting at DOE's Berkeley National Lab ABPDU for Industrial Chemicals, Fuels, Materials, and Food Ingredient Production

Abstract:

Robust and flexible process development, piloting and demonstration is key to scaling up and commercializing new products important for both emerging and established companies in the bioeconomy using fermentation, chemistry, separations, and / or purification processes. The US Dept. of Energyâ€™s (DOE) Advanced Biofuels - *and Bioproducts* - Process Demonstration Unit (ABPDU), located at Lawrence Berkeley National Laboratory (LBNL) in the San Francisco Bay Area is pioneering new collaborations to accelerate this development and de-risk technologies with our industry, academic and National Lab partners. Over the past 3-4 years, our experienced and well-trained team has worked with over 30 start-up and larger companies from around the country and overseas to develop and scale processes in our facility for products as diverse as chemical intermediates, biomaterials, biofuels and proteins to enable process debottlenecking and commercialization using our world-class bench-to-pilot scale process and analytical equipment and software. The ABPDU serves as a collaboration facility where researchers and developers can incubate and improve their processes, demonstrate their technologyâ€™s feasibility, produce kg-scale product volumes for application testing, and create technology transfer packages for scale-up / scale-down prototyping with other facilities. This presentation will highlight several key outcomes from our partnered projects and describe our technical and project management approaches with a focus on industry-friendly IP and cost terms to work alongside our seasoned team and access the facility and its expanding capability and equipment.

Cold model study of hydro-dynamic parameters affecting the performance of re-circulating fluidized bed

*Nitin Lokachari [a,b] *, Sachin Tomar [a], Raman Sharma [c]*

[a] *Department of Chemical Engineering - Birla Institute of Technology and Sciences (BITS), Pilani - 333031, Rajasthan, India.*
[b] *School of chemistry – National University of Ireland, Galway.*
[c] *Senior Scientist, CSIR – National Environmental Engineering Research Institute, Delhi Zonal Centre, New Delhi-110028, India.*

* *E-mail: n.lokachari1@nuigalway.ie*

Abstract

The recent Paris climate conference has set some ambitious targets for the researchers world-wide, to limit the global average temperature rise to well below 2 ^{0}C, which is considered to be the threshold limit before which the catastrophic effects are visible. Development of low carbon technologies are required in-order to meet these objectives as well as to regulate the increasing CO_2 emissions. Chemical Looping Combustion (CLC) is a novel oxy-combustion CO_2 capture technique. In CLC, the fossil fuel combustion is separated into two parts using metal oxide as the catalyst/ bed material. Firstly, the metal oxide is oxidized with air in an air reactor (AR), followed by the reduction of metal oxide using the fuel, in a fuel reactor (FR). To fulfil this criteria of combustion for CLC, some researchers have tested interconnected reactors arrangement i.e., oxidation and reduction in two separate reactors and some have used single reactor which can act as both AR and FR. RCFB is a spouted fluidized bed with a central draft tube to induce higher circulation rates of the bed material. Design and operation of RCFB is simple as it does not contain cyclone separator and loop seal arrangement which are considered as the main sites for heat loss and particle attrition. The hydrodynamic parameters such as pressure drop, operating bed voidage, solid circulation rate, gas by-passing and suspension density and their dependence on the operating parameters such as bed inventory, superficial air velocity, particle size and spacer section is reported.

1. Introduction

Re-circulating fluidized bed is a concentric tube arrangement with inner tube known as riser or draft tube, and the outer tube is known as down-comer. Bed material is fluidized through the draft tube by compressed air and down-comer carries free falling particles from the draft tube. The draft tube induces higher solid circulation rate in the RCFB reactor which will ensure lesser cluster formation [1]. RCFB reactor can be used for applications like combustion and gasification. Interconnected reactor as well as single reactor arrangements can be used for such industrial applications. The design of the RCFB involves use of a freeboard instead of a cyclone separator. As cyclone separator is one of the main sites for attrition, cyclone free arrangement will ensure lesser generation of fine particles [2]. In possible use of RCFB reactor the complex loop-seal arrangement, which requires proper pressure balance around them, is not a part of system will ensure lesser gas bypassing [3]. Heat losses will be less due to the absence of cyclone separator and loop seals. RCFB has been widely used for variety of operations other than combustion and gasification, such as incineration of liquid waste, grain drying, blending and mixing, coating of tablet applications, but has never been tested for chemical looping combustion application [4-6]. Hence, it is important to focus on the hydrodynamics. Chemical looping combustion requires two separate reactor arrangement (Fig. 1), Air reactor (AR), Fuel reactor (FR) and metal oxide as oxygen carrier that circulates from AR to FR [7-10]. Oxidation of metal oxide (equation 1) happens in AR, using atmospheric air as fluidizing medium there by stripping off nitrogen, which is vented out into the atmosphere, followed by reduction (equation 2) in FR, where metal oxide particles carrying oxygen, gets reduced and travel back to AR for the cycle to continue [11-13]. CLC can be carried out in RCFB reactor, where draft tube acts as AR and down-comer as FR.

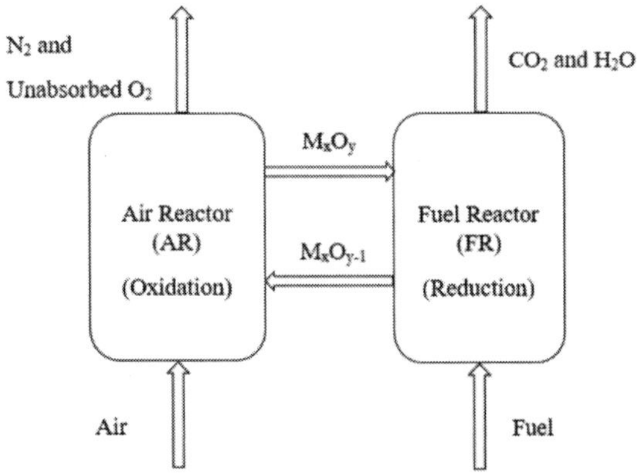

Air Reactor (AR)

$$Me + \frac{1}{2}O_2 \rightarrow MeO \tag{1}$$

Fuel Reactor (FR)

$$\begin{aligned}&(2n+m)MeO + C_nH_{2m} \rightarrow \\ &(2n+m)Me + mH_2O + nCO_2\end{aligned} \tag{2}$$

Fig. 1. Basic reactor arrangement for CLC.

In this paper, cold model study of a RCFB has been carried out to study parameters affecting its performance in different operating conditions. Cold models act as the first set of experiments before real time pilot plant studies. They suggest modifications to be brought into the actual equipment. Cold models mainly focus on studying long term steady operation of a reactor system and also validate hydrodynamic models [14]. Hydrodynamic study is essential for fluidized beds for better understanding of the design and operation and choosing of suitable operating parameters/conditions for a real time operation [15].

2. Materials and Methods

2.1 Experimental setup and methodology

Perspex made transparent, semi-circular re-circulating fluidized bed setup was fabricated for the cold model study at ambient conditions i.e. 1 atm and 300 K. The hydrodynamic data obtained from a semi-circular column are observed to be same to those obtained from a full column [1]. The transparent reactor helps visualize the flow patterns. The cold model setup has four sections; from the bottom, air jet tube section, spacer section, draft tube section and freeboard section. The spacing between the draft-tube and the perforated plate is the spacer section. The picture of the RCFB cold model setup and schematic representation are shown in Fig. 2 & Fig. 3 respectively. Indian Standard (I.S.) sand of different grades (1, 2, 3) was used as the bed material (properties are listed in Table 1) and compressed ambient air for fluidizing the bed particles. As the cold model experiments are multi-phase flow, Geldart's classification insights the particle size distribution and its properties [16].

Sand grade	Geldart's classification	Particle size (mm)	Average particle size (mm)	Bulk density (kg/m³)	Particle density (kg/m³)	Dense bed voidage
1	D	1-2	1.235	1463	2523	0.42
2	B	0.5-1	0.526	1422	2503	0.432
3	B	0.09-0.5	0.30	1250	2500	0.5

Table 1. Characteristics of the bed material (I.S sand).

2.2 Design of experiments

Table 2 lists all operating conditions/variables considered for the experiments, such as air flow rate, spacer section, bed inventory and sand grade (Table 2 & 3).

Variables	Range
Sand grade	1, 2, 3
Bed inventory (kg)	4, 6, 8, 9, 10
Spacer section (cm)	3, 8, 15
Diameter of air jet section (cm)	3

Table 2. Variables considered for the cold model experiments.

Serial no.	Sand grade	Air jet tube diameter (cm)	Spacer section (cm)	Bed inventory (kg)	Repetition for each bed inventory
1	1, 2, 3	3	3	4, 6, 8, 9	5
2	1, 2, 3	3	8	4, 6, 8, 9	5
3	1, 2, 3	3	15	8, 9, 10	5

Table 3. Design of cold model experiments conducted in a RCFB reactor.

2.3 Measurements and calculations

The compressed air velocity (V), pressure drop across the riser (ΔP_r), pressure drop across the down-comer (ΔP_d), pressure drop between the bottom of riser and down-comer (ΔP_{r-d}), particle velocity in the down-comer section (U_{pd}), height of the spout that leaves the draft tube during stable operation (H_{dt}), height of the bed particles in the down-comer section (H_{dc}) were measured simultaneously during the stable operation of the cold model RCFB setup. The operating bed voidage in the riser and down-comer, solid circulation rate, suspension density in the down-comer section and gas bypassing were calculated using the formulae, available in the existing literature as listed in Table 4.

Fig. 2. Lab scale cold model RCFB

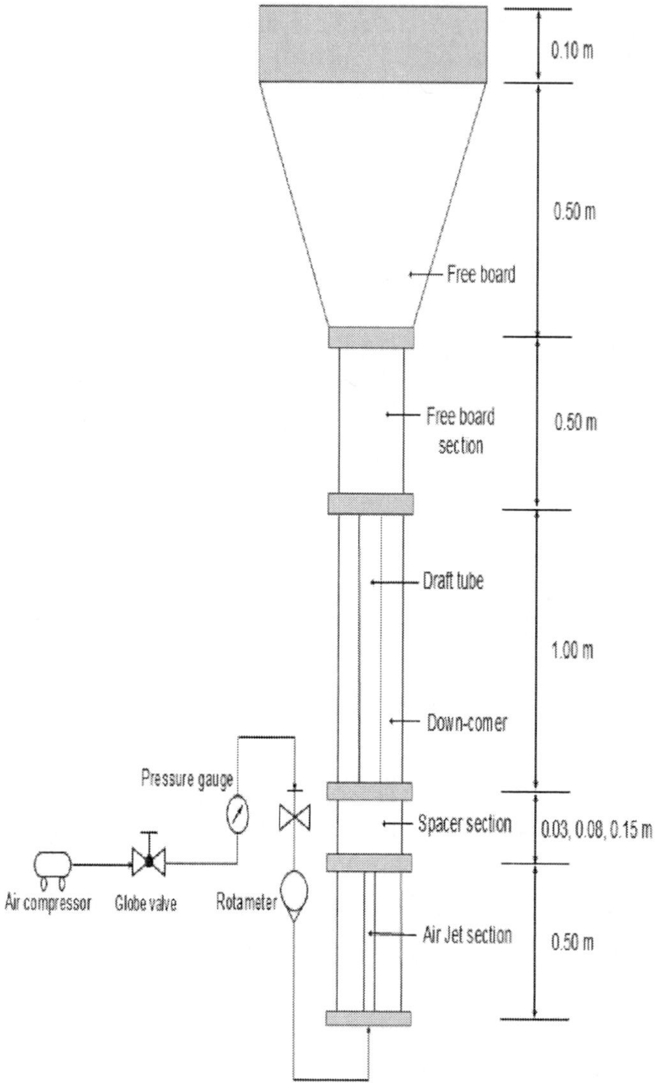

Fig. 3. Schematic representation of

S. no	Parameter (Symbol)	Equation	Literature
1	Operating bed voidage in the riser (e_r)	$e_r = \frac{H_{dt} - H_s(1-e)}{H_{dt}}$	BJ Alappat et al (1996) [17]
2	Operating bed voidage in down-comer (e_d)	$i = H_{dt}(1-e_r)\rho_p A_r + H_{dc}(1-e_d)\rho_p A_d + (H_{dt} - H_{dc})(1-e_r)\rho_p A_r$	BJ Alappat et al (2001) [6]
3	Solid circulation rate (W_s)	$W_s = U_{pr}(1-e_r)\rho_p A_r = U_{pd}(1-e_d)\rho_p A_d$	MK Chandel et al (2006) [19]

4	Suspension density (ρ_{sus})	$\rho_{sus} = \rho_p (1 - e_d)$	XS Wang et al (1996) [18]
5	Gas bypassing	$\frac{\Delta P_d}{H_{dc}} = \frac{150\mu(1-e_d)^2 (U_{pd} + U_{fd})}{e_d^2 d_p^2 \phi_p^2} + \frac{1.75\rho_{air}(1-e_d)(U_{pd} + U_{fd})^2}{e_d\, d_p\, \phi_p}$	WC Yang et al (1978) [1]

Table 4 Calculation of parameters from the available literature.

3. Results and Discussion

In this study, experiments were carried out over all the regimes of fluidization starting from fixed bed to pneumatic conveying depending on the cold model geometry. Parameters such as pressure drop, solid circulation rate, suspension density, operating bed voidage, minimum re-circulating air velocity, gas bypassing and their effects on various operating conditions are reported, but the discussion is limited to fast fluidization regime which greatly influences the performance of a RCFB.

3.1 Pressure drop

Pressure drop gradually increased with superficial air velocity till vigorous bubbling regime. With further increase in the air velocity, fast fluidization regime is visible where bed particles are in continuous stable recirculation from riser to down-comer and pressure drop variation with air velocity is more or less constant in this regime as shown in the Fig. 4 (a). Pressure drop increased with bed inventory and spacer section as shown in the Fig. 4 (b). Increase in the inventory increases the hindrance for the flow and hence pressure drop increases. Since the increase in the spacer section allows more gas into the down-comer section rather than riser, the amount of air available for the riser section is less resulting in higher pressure drop.

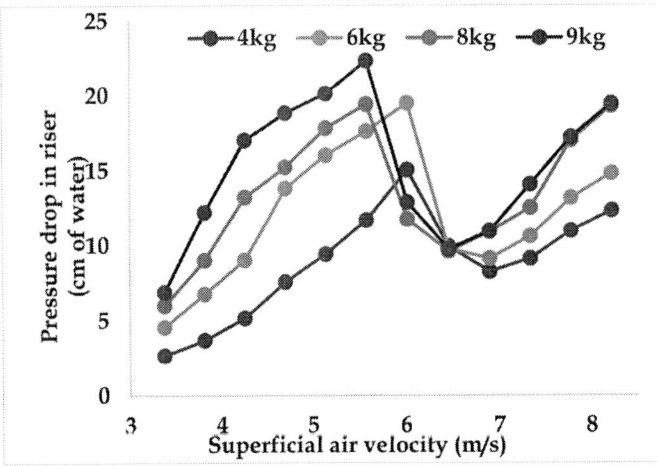

Fig 4 (a). Variation of pressure drop with air velocity for 3 cm spacer section for sand grade 1.

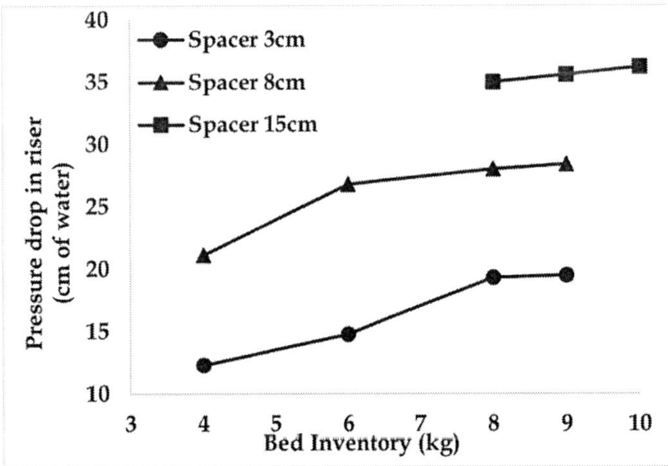

Fig 4 (b). Variation of pressure drop with bed inventory for 3, 8, 15 cm spacer sections for sand grade 1.

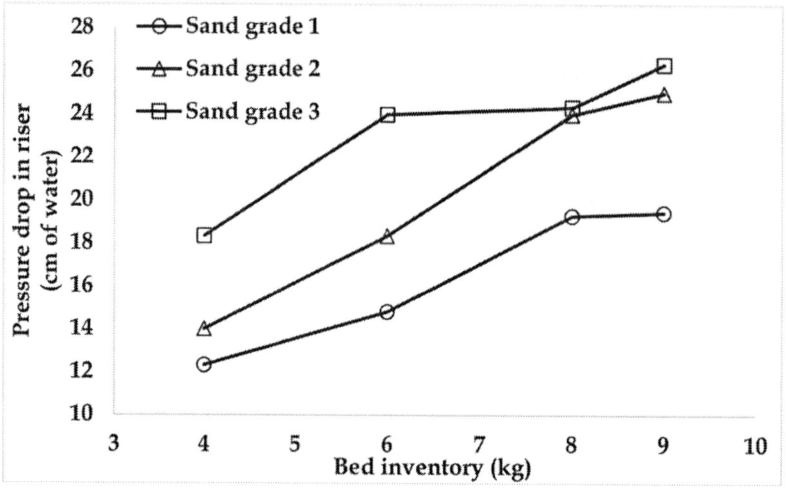

Fig 4 (c). Variation of pressure drop with particle size for 3 cm spacer section for sand grade 1, 2, 3.

3.2 Solid circulation rate

Solid circulation is one of the critical variables that helps maintain the temperature of the system at a desired value and influences heat balance and stable operation of the reactor [20-22]. Solid circulation rate increased with increase in the length of the spacer section for a given grade of sand as shown in Fig. 5 (a) which is due to sufficient fluidizing air velocity to entrain the particles which result in increased entrainment zone. It was also observed as shown in Fig. 5 (b), solid circulation rate increased with the increase in the particle size for a given spacer section. This effect could be attributed to the wall effects as more particles travelling close to the wall results in lower velocity of the particles (in the case of grade 2 and 3) than the particles that are away from the wall (in the case of grade 1) [23]. Solid circulation rate was lesser in the case of sand grade 2 and 3 than in grade 1 because fine particles have more tendency to agglomerate and form clusters [24].

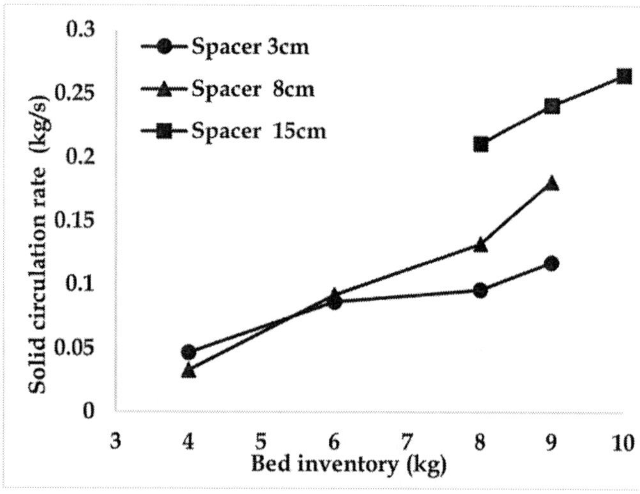

Fig. 5 (a). Variation of solid circulation rate with bed inventory for sand grade 2.

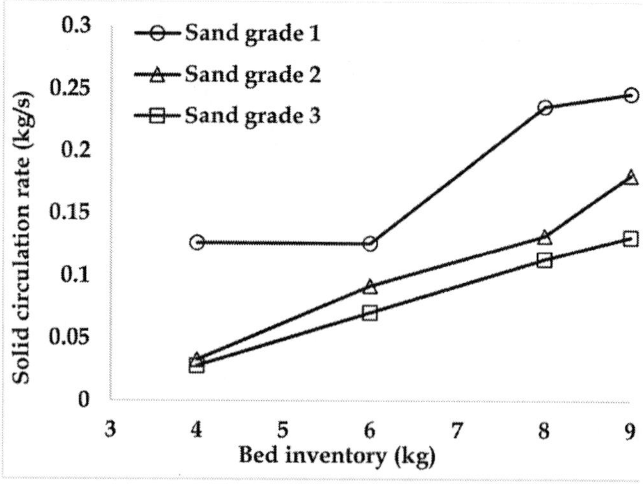

Fig. 5 (b). Variation of solid circulation rate with bed inventory for sand grades 1, 2, 3 for 8 cm spacer

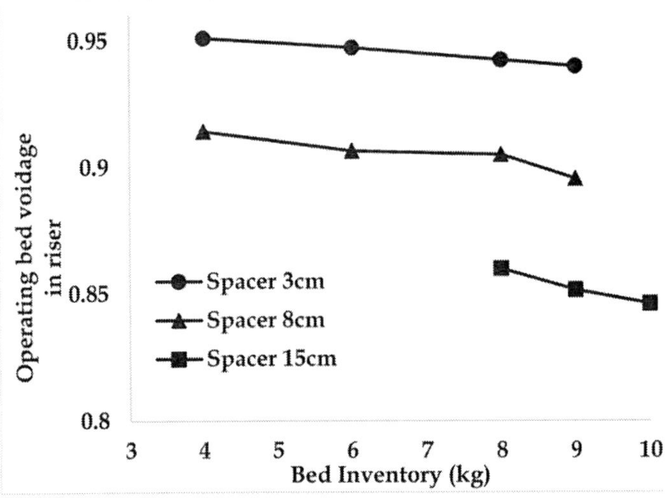

Fig. 6(a). Variation of operating bed voidage in riser with bed inventory for sand grade 2 at 3, 8, 15 cm spacer section.

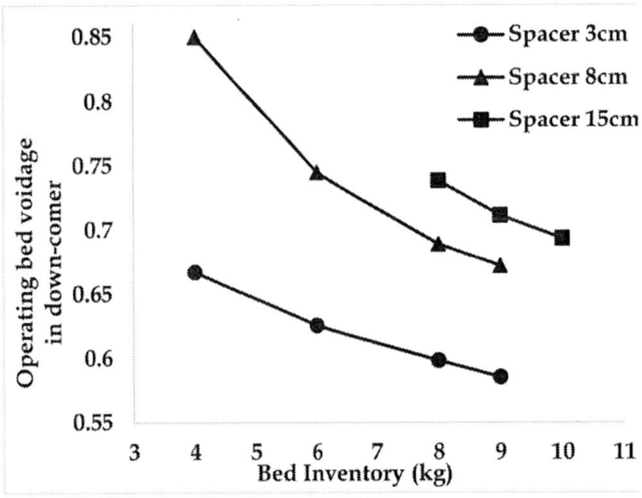

Fig. 6(b). Variation of operating voidage in down-comer with bed inventory for sand grade 2 at 3, 8, 15 cm spacer section.

3.3 Operating bed voidage

Low values of operating bed voidage besides the wall ensures higher rate of bed to wall heat transfer, because the bed material is closely packed [25]. Hence operating bed voidage strongly influences the design and scale up of the reactor. As shown in the Fig. 6 (a) and 6 (b), plotted for sand grade 2, with increase in the spacer section, operating bed voidage followed an exactly opposite trend for riser and down-comer sections. In the riser, operating bed voidage decreased with increase in the spacer section and vice-versa in the down-comer. This is due to the increase in gas bypassing when the spacer section is increased, more amount of air gets into down-comer area which is supposed to enter the riser.

3.4 Suspension density

Bed-to-wall heat transfer, gas-particle residence time distribution and differential pressure fluctuation strongly depend on the suspension density [26]. Suspension density also plays an important role in the design and operation of any fluidized bed system. Suspension density is found to increase with increase in the bed inventory, for a given sand grade, due to the reduction of the operating voidage in the down-comer, Fig. 7 (a). For a given grade of sand, suspension density decreased with increase in the length of the spacer section. It was found that suspension density increased with an increase the particle size for a given spacer section, Fig. 7 (b).

3.5 Minimum re-circulating air velocity

For a particular grade of sand and spacer section, minimum re-circulating air velocity was observed constant with increase in bed inventory, but with increase in the spacer length minimum re-circulating air velocity increased, as depicted in Fig 8.

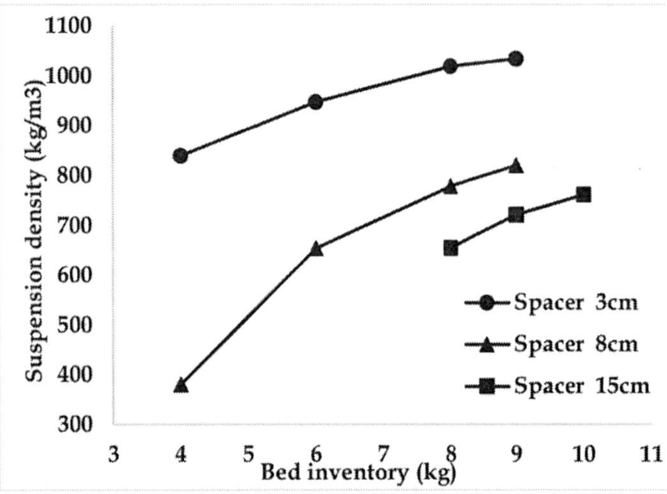

Fig 7(a): Variation of suspension density with bed inventory for 3, 8, 15 cm spacer section for sand grade 2.

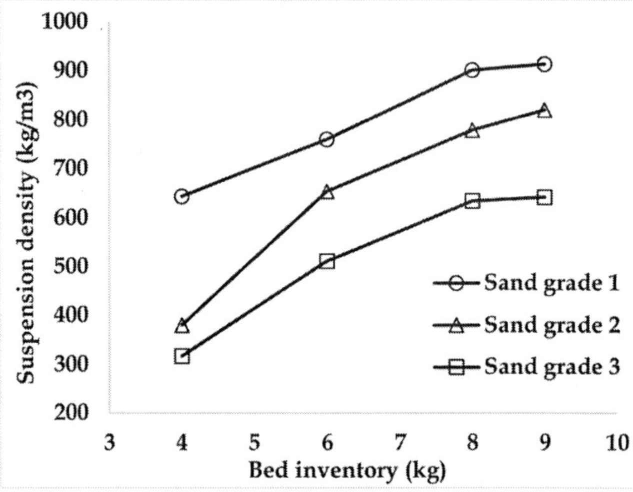

Fig 7(b): Variation of suspension density with bed inventory for 8 cm spacer section for sand grades 1, 2, 3.

3.6 Gas bypassing

Gas bypassing is not a desired phenomenon in any fluidized bed operation, as it affects the flow behavior and gas residence time which is one of the important aspects for the success of the circulating fluidized bed [27-28]. In our study, gas bypassing decreased with an increase in the bed inventory, Fig. 9 (a), due to the fact that higher bed inventory increases the resistance in down-comer which resists air flow towards the down-comer. Gas bypassing increased with increase in the length of the spacer section, due to the fact that air starts escaping towards the down-comer. It was also found that, gas bypassing decreased with the decrease of particle size for a given spacer section, Fig. 9 (b).

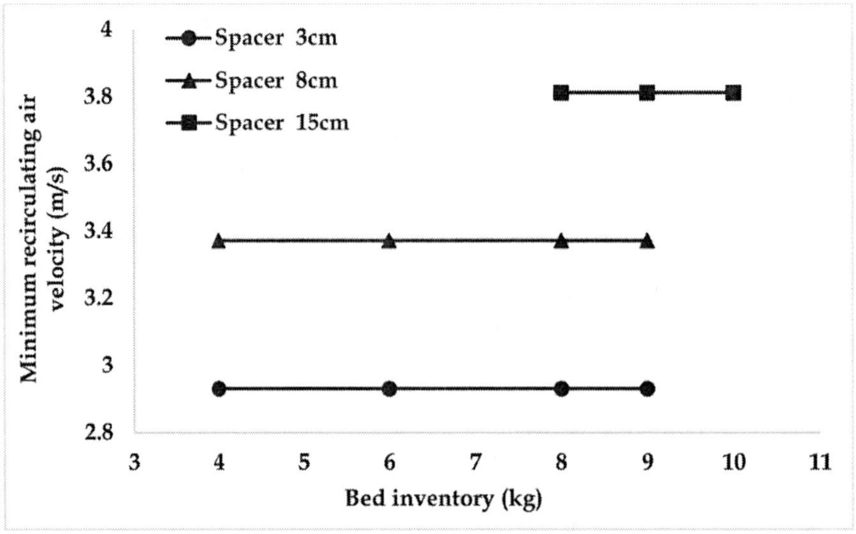

Fig. 8. Variation of minimum re-circulating air velocity with bed inventory for 3, 8, 15cm spacer section.

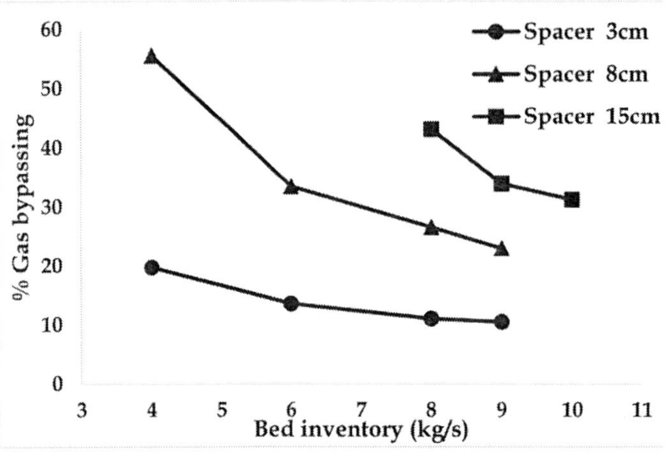

Fig. 9(a). Variation of gas bypassing with bed inventory for sand grade 2.

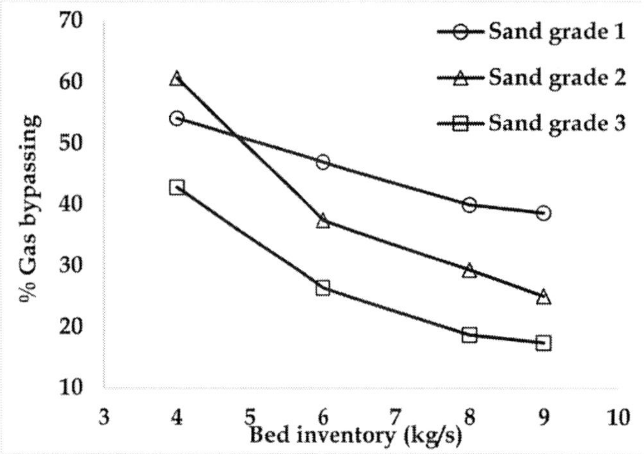

Fig. 9(b). Variation of gas bypassing with bed inventory for 8 cm spacer section for sand grade 1,

4. Conclusions

In the current work, hydrodynamics of a Re-circulating fluidized bed cold model setup has been studied at ambient conditions. Parameters that affect the performance of a RCFB like pressure drop, solid circulation rate, suspension density, operating bed voidage, minimum re-circulating air velocity and their effects on various operating conditions have been reported and justified. It was found that all the parameters depend on bed inventory, spacer section and particle size. The minimum re-circulating air velocity increased with the spacer section and found to be independent of bed inventory.

With increase in bed inventory, pressure drop increased as expected and there is also an increase in solid circulation rate and suspension density, on contrary operating bed voidage and gas bypassing decreased. With an increase in the spacer section, it was found that pressure drop, solid circulation rate, operating bed voidage in the down-comer and gas by-passing increased, whereas suspension density and operating bed voidage in riser are decreased. Likewise with an increase in particle size, solid circulation rate and suspension density increased, gas by-passing decreased. Such studies are required for the usability of RCFB reactor as a combustor, gasifier and incinerator or chemical looping combustion. Depending on the requirement, the presented data can be used to operate the RCFB reactor for various applications.

A high solid circulation rate and suspension density and low operating bed voidage in the down-comer are necessary criteria for CLC. However cold models are useful as the first set of experiments, furture research should be directed towards real time combustion studies for CLC application in a RCFB reactor.

References

[1] W.C. Yang, D.L. Keairns, Design of recirculating fluidized beds for commercial application, *A.I.Ch.E. Symposium Series,* 74(176) (1978) 218-228.
[2] J. Werther, J. Reppenhagen, Catalyst attrition in cyclones, *Powder Technology*, 111 (2000) 55–69.
[3] P. Basu, L. Cheng, An analysis of loop seal operation in a circulating fluidized bed, *Transactions of the Institution of Chemical Engineers*, 78(A) (2000) 991–998.

[4] W.C. Yang, D.L. Keairns, Recirculating fluidized bed reactor data utilizing a two dimensional cold model, *A.I.Ch.E. Symposium Series*, 70(141) (1974) 27–40.

[5] B.J. Alappat, V.C. Rane, Performance prediction of a RCFB incineration system, *Journal of Energy Engineering*, 126(2) (2000) 53–65.

[6] B.J. Alappat, V.C. Rane, Solid circulation rate in recirculating fluidized bed, *Journal of Energy Engineering*, 127 (2001) 51-68.

[7] E. Johansson, A. Lyngfelt, T. Mattisson, F. Johnsson, Gas leakage measurements in a cold model of an interconnected fluidized bed for chemical-looping combustion, *Powder Technology,* 134 (2003) 210–217.

[8] E. Kimball, A. Lambert, A. Fossdal, R. Leenman, E. Comte, W.A.P. van den Bos, R. Blom, Reactor choices for chemical looping combustion (CLC) - Dependencies on materials characteristics, *Energy Procedia,* 37 (2013) 567–574.

[9] S. Penthor, K. Mayer, S. Kern, H. Kitzler, D. Wöss, T. Pröll , H. Hofbauer, Chemical-looping combustion of raw syngas from biomass steam gasification – Coupled operation of two dual fluidized bed pilot plants, *Fuel,* 127 (2014) 178–185.

[10] W. Lewis, E. Gilliland, W. Sweeney, Gasification of carbon: metal oxides in a fluidized powder bed, *Chemical Engineering Progress*, 47(5) (1951) 251–256.

[11] A. Lyngfelt, B. Leckner, T. Mattisson, A fluidized bed combustion process with inherent CO2 separation; application of chemical-looping combustion, *Chemical Engineering Science*, 56 (2001) 3101–3113.

[12] H. Leion, T. Mattisson, A. Lyngfelt, Solid fuels in chemical looping combustion, *International Journal of Greenhouse Gas Control*, 2 (2008) 180-193.

[13] T. Mattisson, G.F. Labiano, B. Kronberger, A. Lyngfelt,, J. Adánez, H. Hofbauer, Chemical looping combustion using syngas as fuel, *International Journal of Greenhouse Gas Control*, 1 (2007) 158–169.

[14] A.R. Bidwe, C. Hawthorne, Yu Xizhi, H. Dieter, G. Scheffknecht, Cold model study of a dual fluidized bed system for the gasification of solid fuels, *Fuel,* 127 (2014) 151–160.

[15] M. Hamzehei, Study of Heat Transfer and Hydrodynamics in the Fluidized Bed Reactors, Heat Transfer - Mathematical Modelling, Numerical Methods and Information Technology, Prof. Aziz Belmiloudi (Ed.), ISBN: (2011) 978-953-307-550-1, InTech, Available from: http://www.intechopen.com/books/heat-transfermathematical-modelling-numerical-methods and information-technology/study-of-heat-transfer-andhydrodynamics-in-the-fluidized-bed-reactors.

[16] D. Geldart, Types of gas fluidization, *Powder Technology,* 7 (1973) 185–195.

[17] B.J. Alappat, V.C. Rane, Studies on a recirculating fluidized bed incinerator cold model, *International Journal of Energy Research*, 20(4) (1996) 313–326.

[18] X.S. Wang, B.M. Gibbs, M.J. Rhodes, D. Geldart, A Correlation for the Suspension-to-Wall Heat-Transfer Coefficient in Circulating Fluidized Beds, *Industrial and Engineering Chemistry Research*, *35*(10) (1996) 3822–3823.

[19] M.K. Chandel, B.J. Alappat, Pressure drop and gas bypassing in recirculating fluidized beds, *Chemical Engineering Science,* 61(5) (2006) 1489-1499.

[20] B. Kronberger, E. Johansson, G. Löffler, T. Mattisson, A. Lyngfelt, H. Hofbauer, A Two-Compartment fluidized Bed Reactor for CO_2 Capture by Chemical-Looping Combustion, *Chemical Engineering & Technology,* 27(12) (2004) 1318-1326.

[21] A. Abad, J. Adánez, F.G. Labiano, L.F. de Diego, P. Gayán, J. Celaya, Mapping of the range of operational conditions for Cu-, Fe-, and Ni-based oxygen carriers in chemical-looping combustion, *Chemical Engineering Science*, 62 (2007) 533–549.

[22] F. Wei, Y. Cheng, Y. Jin, Y. Zhiqing, Axial and Lateral Dispersion of Fine Particles in a Bina G-Solid Riser, *The Canadian Journal of Chemical Engineering, 76 (*1998) 19-28.

[23] W.C. Yang, D.L. Keairns, Comparison of recirculating fluidized bed performance in two-dimensional and three-dimensional beds, *Fluidization Technology*, 2 (1976) 51–63.

[24] A.K. Sharma, K. Tuzla, J. Matsen, J.C. Chen, Parametric Effects of Particle Size and Gas Velocity on Cluster Characteristics in Fast Fluidized Beds, *Powder Technology*, 111 (2000) 114-122.

[25] D. Shi, R. Nicolai, L. Reh, Wall-to-Bed Heat Transfer in Circulating Fluidized Beds, *Chemical Engineering and Processing: Process Intensification*, 37 (1998) 287-293.

[26] A.S. Issangya, D. Bai, H.T. Bi, K.S. Lim, J. Zhu, J.R. Grace, Suspension densities in a high-density circulating fluidized bed riser, *Chemical Engineering Science*, 54 (1999) 5451-5460.

[27] H.H. Shih, C.Y Chu, S.J. Hwang, Solids Circulation and Attrition Rates and Gas Bypassing in an Internally Circulating Fluidized Bed, *Industrial and Engineering Chemistry Research*, 42 (2003) 5915-5923.

[28] H.S. Ahn, W.J. Lee, S.D. Kim, B.H. Song, Solid Circulation and Gas Bypassing in an Internally Circulated fluidized Bed with an Orifice-Type Draft Tube, *The Korean Journal of Chemical Engineering*, 16(5) (1999) 618-623.

502136 Troubleshooting Anomalous Behaviour in a Solid-State Reaction Process

Abstract:

A novel solid-state process for increasing molecular weight of a polymer powder was scaled-up to 3 m3 tumble dryer before going to full-scale production. The pilot plant process occasionally showed damaged internals, which eventually ended in serious damage to the reactor. A full-blown root cause analysis, including analysis of the powder properties before and after the incidents, did not reveal an explanation. Only after further detailed analysis of the plant signals around the incident and an understanding of the underlying physics of the process, the root cause was revealed. The full-scale process was started up with operating parameters based on the new findings and still runs smoothly today. However, a one-time error in operation of the full-scale process indeed reproduced bent internals, proving that the solution found is at the heart of the problem and that it protects the full-scale process from damage and mechanical failure.

498543 Production of Chemicals from the Thermal Cracking of Castor Oil, Castor Oil Methyl Ester and Ricinoleic Acid

Abstract:

This work evaluates the experimental results of three different processes to obtain heptaldehyde, undecylenic acid and methyl undecylenate through castor oil, castor oil methyl ester, and ricinoleic acid. Heptaldehyde and undecylenic acid are important chemicals on food, packaging, ink, solvent and polymers industries. Methyl undecylenate is commonly used as a Flavoring agent and also an important route to obtain undecylenic acid. Methyl undecylenate was obtained by the thermal cracking of castor oil methyl ester. In order to obtain heptaldehyde, methyl undecylenate, and undecylenic acid, the thermal cracking was performed using commercial castor oil, castor oil methyl ester (obtained by castor oil transesterification reaction) and commercial ricinoleic acid, respectively. Thermal cracking temperatures were 530, 545, 560 and 575 °C with 400 g/h mass flow on the feed. All materials were blended with distilled water, considering the same proportion (75 % material, 25 % water). Temperature influence on bio-oil yields was observed, likewise different behaviors to each material. Iodine index, acid index, and density were determined to all bio-oil samples and their content were analyzed by GC-FID. The best condition to obtain higher yields of each chemical were pointed. With regard to the castor oil, higher yields of heptaldehyde and undecylenic acid were found on experiments at 530 °C. The same temperatures resulted in higher yields concerning the ricinoleic acid. The cracking of methyl ricinoleate results on higher yields of heptaldehyde at 530 °C and methyl undecylenate at 560 °C.

NEEM COATED UREA- A FISCAL APPROACH FOR FARMERS

Sara Ahsan, Ali Ayub

Abstract

Soil fertility is determined by three major elements namely Nitrogen, Phosphorus & Potassium (N, P, K) of which nitrogen plays a very important role. For this reason, Urea (Containing 46% of N) consumption is very high all over the world. Generally, the plants are able to recover a fraction of Nitrogen present in Urea and this fraction is known as Nitrogen Use Efficiency (NUE). This NUE stands anywhere between 30 to 50%. Thus, two third of Urea Nitrogen escapes from the soil in the form of nitrogen, ammonia and nitrates etc. and is not used by the plants. Engro Fertilizers, Pakistan, researched and experimented use of "Azadirachta indica" commonly known as neem oil as nitrification inhibiting agent. Based upon the results of extensive laboratory trials, coating urea with neem helps in slow release of nitrogen, nourishing the crops for a longer period. The process reduces pollution of groundwater. There is an increase in crop yield and also serves as an efficient natural insecticide. It also serves as an anti-caking agent and will reduce the amount of urea required by a plant by ~10-20%.

This paper explicates Engro's experience and research on neem coated urea, methods of preparing an effective neem oil emulsion, quantities of raw materials to be used per unit of urea, its uses and results obtained after extensive laboratory and commercial trials.

INTRODUCTION

Snam Progetti divides Urea plant into following major sections i.e.

1. Urea synthesis and NH3, CO2 recovery at high pressure.
2. Urea purification and NH3, CO2 recovery at medium and low pressures.
3. Urea concentration.
4. Urea prilling.
5. Waste water treatment.

Snamprogetti ammonia stripping process is characterized by a urea synthesis loop operating at about 160 Kg/cm2 (abs); with ammonia to carbon dioxide molar ratio at urea reactor inlet of 3.3 to 3.6. This allows a high CO2 conversion into urea in the reactor itself. There are two kinds of chemical reactions at the same time in the urea reactor:

$$\mathbf{2NH_3 + CO_2 \leftrightarrow NH_2\text{-}COO\text{-}NH_4 + 32560\ kcal/kmol\ of\ carbamate\ (at\ 1\ atm;\ 25°C)}$$

$$\mathbf{NH_2\text{-}COO\text{-}NH_4 \leftrightarrow NH_2\text{-}CO\text{-}NH_2 + H_2O - 4{,}200\ kcal/kmol\ of\ urea\ (at\ 1\ atm;\ 25°C)}$$

First reaction is very fast & exothermic and the second one is weakly endothermic and occurs in the liquid phase at low speed. Downstream the urea synthesis, the decomposition and relevant recovery of unconverted chemical reagents occurs. It is carried out in three subsequent steps: High Pressure Decomposition in H.P. stripper; Medium pressure Decomposition in M.P. Decomposer and, finally, Low Pressure Decomposition in Low Pressure Decomposer.

The decomposition reaction is the reverse reaction of the first one above showed, viz.:

$$\mathbf{NH_2\text{-}COO\text{-}NH_4 \leftrightarrow 2NH_3 + CO_2 - Heat,}$$

And, as can be inferred from the equation, it is promoted by reducing pressure and/or adding heat.

The urea reactor effluent solution enters the stripper, under slightly lower pressure than the urea reactor, where a fair part of the unconverted carbamate is decomposed in NH3 and CO2. The overall yield of the H.P. synthesis loop referred to CO2 is as high as 83-85% (on molar basis). Downstream of the stripper residual carbamate and ammonia are recovered in two recycle stages operating at 18.5 kg/cm2 (abs) (namely M.P. section) and 4.8 kg/cm2 (abs) (namely L.P. section) respectively.

Ammonia and carbon dioxide vapors from the stripper top, after mixing with the carbonate recycle solution from M.P. section, are condensed, at the same pressure level of the stripper itself, in the 1st and 2nd High Pressure carbamate condensers, thus producing the L.M.P and the L.P. steam which are used in downstream sections. After separating the inert gases which are passed to M.P. section, the carbamate solution is finally recycled to the reactor bottom by means of a liquid/liquid ejector, which exploits H.P. ammonia feed to reactors as motive fluid.

Downstream of the two recovery stages urea is concentrated to 99.75% via two vacuum concentrator stages. The solution leaving the L.P. decomposer bottom with about 70% wt. urea, is sent first to the tube of vacuum pre concentrator and then pumped to first vacuum concentrator both operating at a pressure of 0.34 kg/cm2 (abs). The mixed phase coming out from the process side of the first vacuum concentrator enters the first gas-liquid vacuum separator from where vapors are again extracted by the first vacuum system while the urea melt (~ 95 wt%), by gravity & differential pressure flow, enters the bottom of second vacuum concentrator operating at a pressure of 0.03 kg/cm2 (abs). The extracted urea melt (~99.75% pure) from the second concentrator is sent to prilling tower. Droplets of molten urea from the prilling bucket fall along the natural draught prilling tower and solidify and cool while encountering a countercurrent air flow.

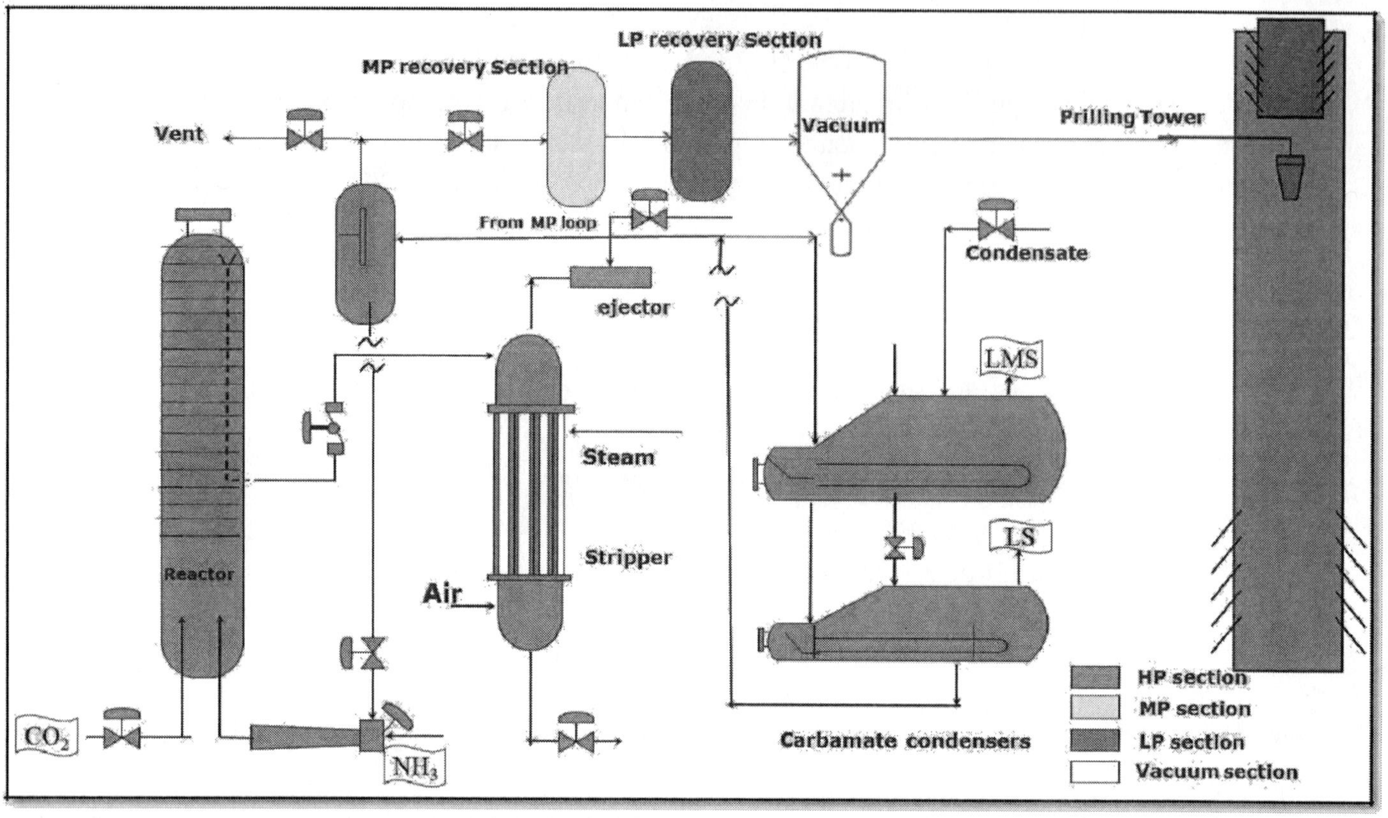

Figure 1. Process Description: Picture elucidating a short process description of Snam's urea plant.

NEEM COATING ON UREA PRODUCT- AN EXPERIENCE AT ENGRO

Purpose of coating Neem on Urea Product:

With 46% N content, Urea is the most popular Nitrogenous fertilizer around the world. When Urea is applied to soil, it is first hydrolyzed into Ammonical form (Ammonium Ion i.e. NH4+) and then to nitrite (NO2-), followed by to nitrate (NO3-) forms by the process called nitrification. Nitrification process is necessary for making nitrogen available to plants. However, if the process of nitrification is too rapid; nitrogen will escape to atmosphere and plants will not be able to recover it from Urea efficiently. Generally, the plants are able to recover a fraction of Nitrogen present in Urea and this fraction is known as Nitrogen Use Efficiency (NUE). This NUE stands anywhere between 30 to 50%. Thus, two third of Urea Nitrogen escapes from the soil and is not used by the plants. Hence this lead us to regulate the urea hydrolysis and nitrification. This regulation is done by some chemical/natural agents called Nitrification inhibitors. However, most of the nitrification inhibitors are costly chemicals (examples nitrapyrin, dicyandiamide and ammonium thiosulphate) which are not economically feasible.

It has been scientifically established that Neem oil serves as an effective nitrification inhibitor whose benefits when coated on urea area mentioned below:

- Neem coating leads to more gradual release of urea, helping plants gain more nutrients and resulting in higher yields.

- Lower underground water contamination due to leaching of urea.
- Neem serves as a natural insecticide and an anticaking agent

Properties of Neem Oil:

The properties of neem oil used are as follows:

1. Neem Oil purity = 98.5%+
2. Color= Yellowish brown clear liquid
3. Moisture = 0.5% Max.
4. Specific gravity = 0.9 to 0.94 at 30 oC
5. Iodine value = 65 to 80
6. Azadirachtin Content = min 150 ppm
7. Saponification Value = 180 to 205
8. Up saponi cable matter = < 2 %
9. Density of neem oil = $0.92 g/cm^3$
10. Viscosity of neem oil = 35CSt

As per standard limits of PSQA, Azadirachtin content in final product must be greater than 50 ppm corresponding to 250ppm of neem oil coat. Engro started its coating setting a standard in Pakistan of 500ppm of neem oil coat. So, 0.5kg of neem oil is required per ton of urea. Cost of neem oil is 2.21$ per kg.

Process employed for coating of Prilled Urea

Process Description:

Coating of neem oil on urea product requires preparation of solution of neem oil in water. Neem oil is not soluble in water so it requires an emulsification agent. On Laboratory scale, tests were done on the use of Lab grade 25% Ammonical water as an emulsifying agent. The results were favorable for the use of Ammonical water as an agent. It was then decided to use it in our emulsion preparation method.

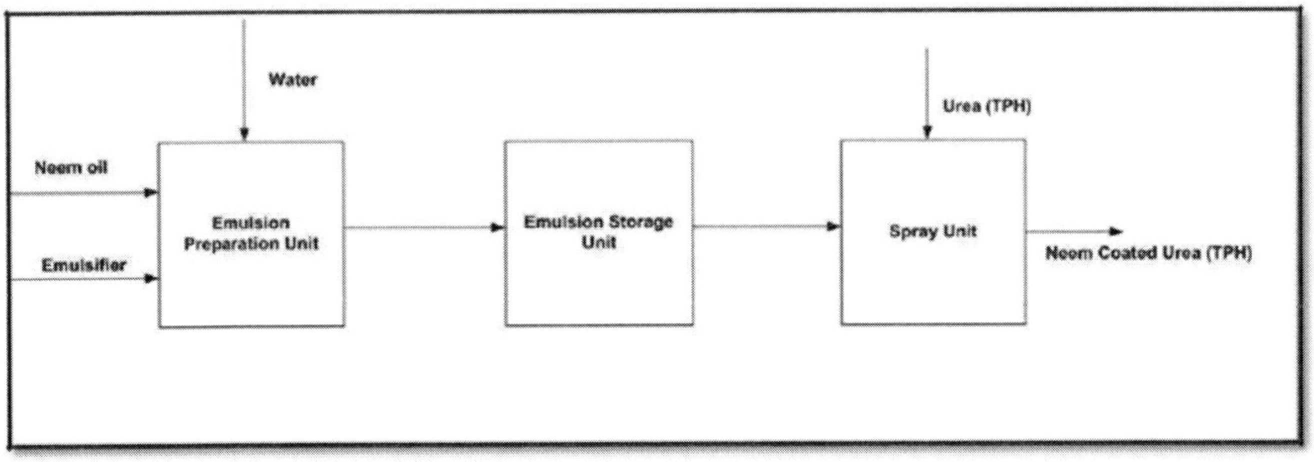

Figure 2: Block Diagram of emulsion preparation process

Below two images depicts neem solution change in color before and after addition of emulsifying agent.

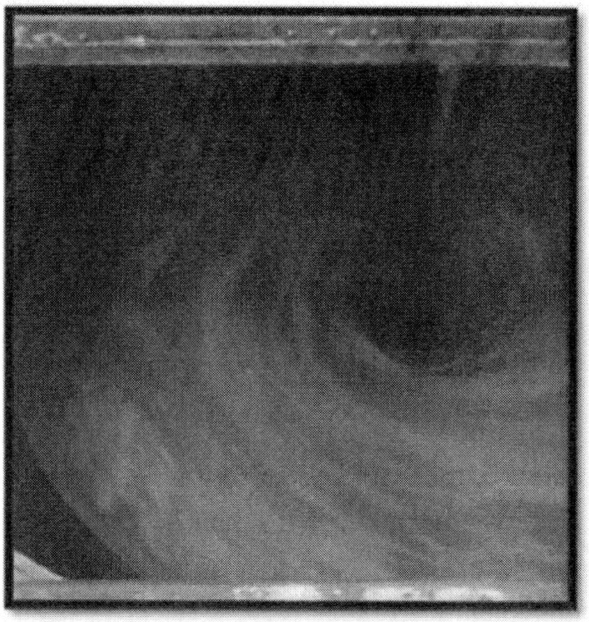

Figure 3: Neem Oil & water (Before addition of Ammonical solution)

Figure 4: Neem emulsion (After addition of Ammonical solution)

Design capacity of Engro's EnVen plant is 3835 MTPD which is equivalent to 160TPH of product over the belt. For coating 500 ppm of neem oil over this product, first and the foremost step was to add accurate quantities of raw materials and emulsifying agent. For the aforementioned urea production, total neem emulsion required was 3835kg, with pure neem oil amount of 1917.5kg (50% of total emulsion). 0.1grams of 25% Ammonical solution is required per 50grams of pure neem oil used and the remaining being water.

Engro's Laboratory researched and registered its test for determining amount of Azadirachtin in the coated product. Below is the Process flow diagram of the facility used during coating activity

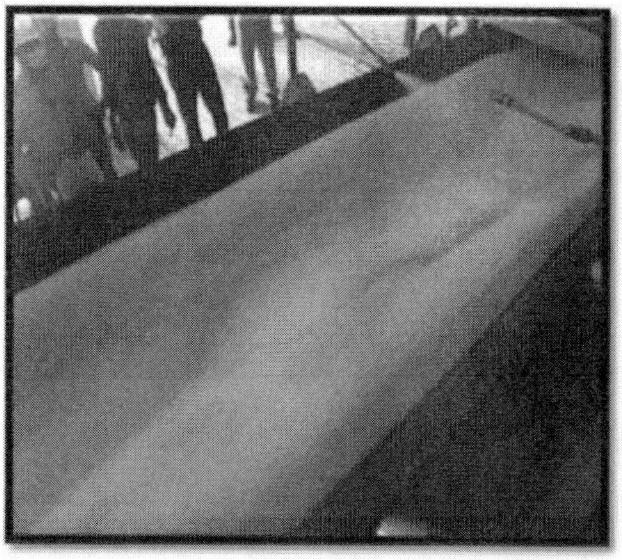

Figure 5: Coated Urea on Conveyer Belt

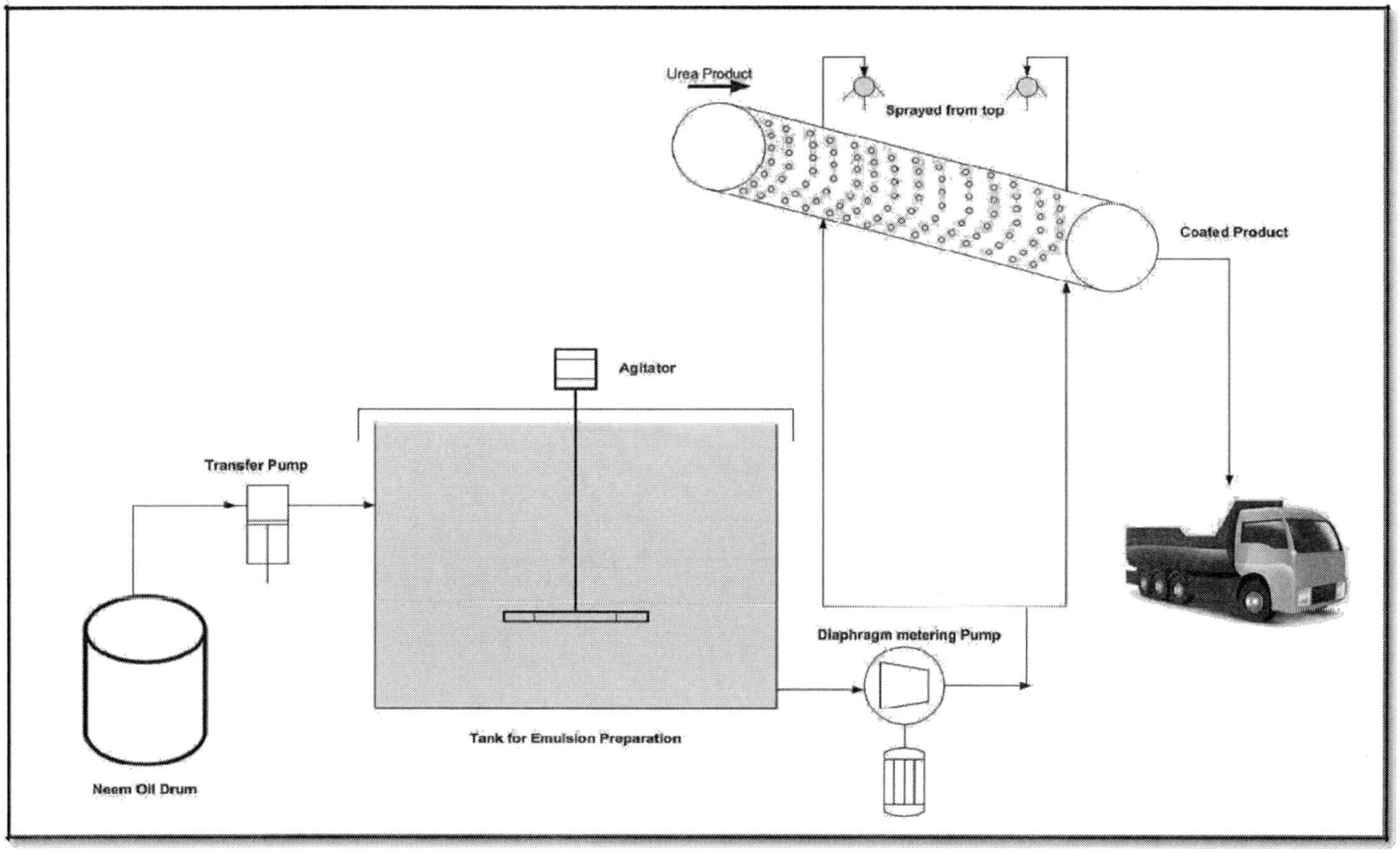

Figure 6: Block Diagram of emulsion preparation process

Before stating coating, stroke testing of the diaphragm metering pump was done. This testing was made basis for selection of the pump strokes during neem coating. The stroke testing results are as mentioned below:

Table 1. Pump stroke testing results

Strokes	Flow (Litre/min)	Flow (Litre/hr)
10	0.5	30
20	0.980	58.8
30	1.37	82.2
40	1.87	112.2
50	2.35	141
60	2.6	156
70	3.15	189
80	3.48	208.8
90	3.9	234
100	4.3	258

For a coat of 500 ppm corresponding strokes needs to be adjusted at 70%. To determine efficaciousness of the spraying unit, we set the strokes primarily at 70% and collected a sample then adjusted them to 50% and collected another sample. These samples reported Azadirachtin content in them to be 99.5ppm and 90ppm. Minimum limit of the active agent was 50ppm, so, strokes were further reduced to 45% and coating as started. Following are the results obtained during coating.

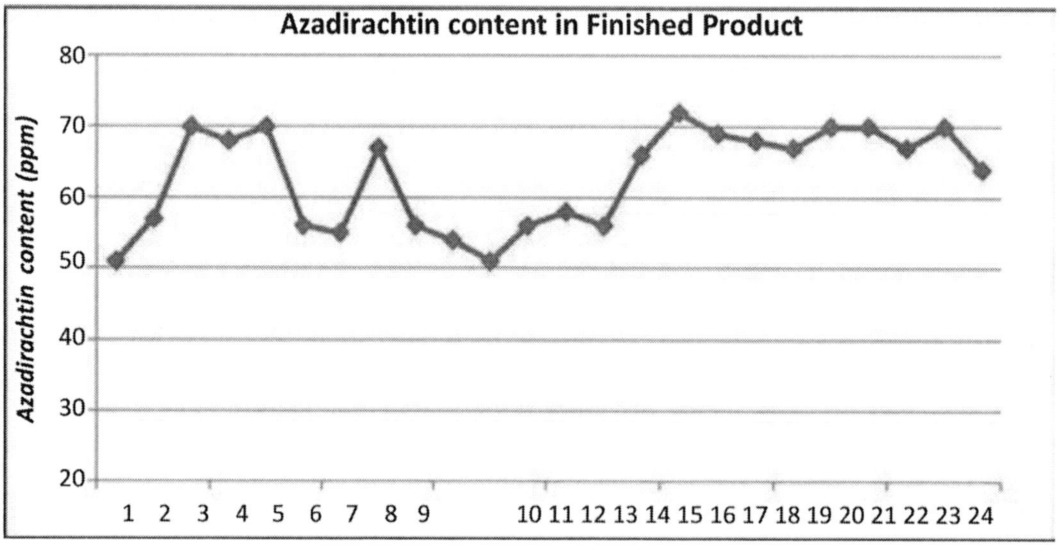

Graphical representation 1: Azadirachtin content in Finished Product

Required Equipment

- One transfer Pump for transferring neem oil into the tank where emulsion is to be prepared (Capacity of used pump: $4m^3/hr$)
- A tank of capacity depending on the design capacity of plant (Tons per hour of product passing through belt conveyer) (Capacity of used Tank: $4.82m^3$)
- Agitator for mixing and emulsion preparation (Rpm of Used agitator: 70)
- Neem emulsion storage tank (Capacity of used Tank: $4.82\ m^3$)
- Gear Pump for transferring neem emulsion to nozzles (Capacity of gear pump used: 220LPH)
- Spray nozzles for spraying neem emulsion on urea over the belt conveyer. (Capacity of each nozzle: 1.6LPM)

Neem in Pakistani Market:

Neem story is very different in Pakistan. Its oil extraction is being done in the country but at a small scale. Neem seed and oil are not traded in Pakistani commodity market. As this tree seed and leaves have toxic properties, so it is not fed to the animals, hence each year all its fruit and leaves are wasted. Consequent result of less production of oil is high prices. After lots of struggles, a pharmaceutical company agreed on selling at 9$ per kg of neem oil. But purchasing at this rate was neither economically feasible nor gives a positive message to the market. Hence, if we want to start neem coating in Pakistan permanently, either its market needs to be improved or else oil should be imported from India or any other country where prices of neem oil are low.

In case we are successful in marketing this Neem coated Urea, it will start a total new industry in Pakistan with lots of parties setting up efficient Neem oil extraction plants and lively hood to thousands of people.

Discussion and Benefits of neem coating on Urea

The decrease in the local urea sales whereby increasing the sale of imported urea led local fertilizer producers to slash their prices. Consequently, to deplete our urea inventory price was lowered by PKR 70. It is important to mention here that government, in Fiscal Year 2016-17 (FY17) budget, reduced urea prices by PKR 400/bag to PKR 1,400/bag. But this price cutting didn't result in the expected sales. So, Engro decided of neem coating on Urea Product, hoping to attract the customers with its uses. Indian government decision of coating 100% produced urea further strengthened our decision. For test run purposes, 10,000 MeT of urea was coated and sent to selected warehouses for marketing.

Engro Fertilizers Limited started first ever coating in Pakistan and homogenized the procedures for production of Neem Coated Urea. To have a uniform and consistent coating of Neem emulsion on urea prills and maintaining the contents of Azadirachtin in final product, Laboratory results and tests were performed after every two hours.

Summarizing the benefits of neem coating on urea product as under:

- Increase in price of coated urea per 50kg bag
- Decrease in nitrification and consequent increase in the yield of crops thus strengthening agricultural business
- Reduction in raw material consumption due to decrease in the amount of fertilizer required by plants
- Reduction in environmental pollution of ground water due to leaching of nitrates and gaseous emissions
- Lively hood to thousands of people
- Profit to the farmers due to reduction in the amount of fertilizers required per crop by approximately 30%
- Neem serves as a natural insecticide

496747 Technoeconomic Assessment of Two-Step Process for Upgrading and Utilizing Used Railroad Ties in an Integrated Biorefinery

Abstract:

In the USA, over 21 million wood railroad ties are produced per year to replace the approximately 20 million used ties that have provided excellent primary service in railroad track over 40 years. Due to the presence of carcinogenic preservatives, these used ties will primarily be either landfilled or incinerated as they do not meet U.S. Environmental Protection Agency (U.S. EPA) stringent requirements. Consequently, a two-step pyrolysis process was developed to upgrade used ties by first extracting wood preservatives (Step One) and converting upgraded used ties into biofuels, bioproducts, power, and heat (Step Two). This study evaluates the techno-economic viability of the proposed two-step process embedded in integrated pyrolysis biorefinery with liquid fuel, power and recovered preservatives as product streams.

The technoeconomic analysis assumed 2.33 million metric tons of used railroad ties are available yearly according to statistics by the Railroad tie association. This analysis considers 3 plants spread across the U.S. to minimize transportation cost with 2358 tons/day throughput each. The rail road ties biorefinery consists of eight areas including preprocessing (A100), thermal desorption (A200) and fast pyrolysis and vapor quenching (A300), hydroprocessing (A400), hydrogen generation (A500), power generation (A600), cooling and other utilities (A700) and waste management (A800). The minimum selling prices as well as sensitivity analyses were performed to determine the impact of thermal desorption conditions on the overall economic.

496556 Valorisation of Three-Phase Olive Mill Wastewater with the Addition of High Protein Co-Substrates

Abstract:

Introduction

Nowadays national and international legislation attempt to enforce reductions in greenhouse gas emissions, while minimization of dependency to fossil fuel gathers all-around agreement. At the same time optimization of waste management strategies towards zero waste generation is a major challenge for modern society. Olive Mill effluents (OME) is the effluent generated during the production of olive oil and accounts for the 60% of the overall weight of the olive fruit. OME is rich in carbon and phenols while its nitrogen content together with the pH are very low rendering common treatment methods ineffective to tackle it. A valorization process that can be used for the management of OME is Anaerobic Digestion (AD). A key attribute of energy efficiency schemes has been the utilization of negative value substrates for biogas generation in anaerobic digesters. Anaerobic digestion (AD) is a biological method for the treatment and valorization of organic material. Its product can be recovered and utilized for heat and/or electricity generation. The efficiency of the process is affected by the operational conditions selected for the digesters and it depends highly on the type of the organic substrate used as feedstock.

Common organic substrates for AD systems include municipal solid waste, agroindustial byproducts and waste including slaughterhouse and dairy industry waste. Several types of AD feedstock are rich in fat, oil and grease (FOG). This is a feature that is highly desirable, as fat has the highest biomethane potential of all natural macromolecules. However, high fat content can inhibit the digesters through flotation, accumulation, pH reduction, and a number of intracellular inhibition processes. Furthermore, agroindustial wastewater, including olive mill effluents (OME), is characterized by high concentration of phenols, low nitrogen availability and low pH. All of the above are known to inhibit the digestion process especially in overloaded or unstable systems.

Aim of this work was to investigate the effect of external addition of nitrogen into digesters operating with OME as single substrate. The co-substrates evaluated were low-value byproducts of animal and plant origin, namely bone and meat meal, blood meal, feather meal, mixture of feather and blood meal, fish meal, soy been meal and maize gluten.

Material and methods

OME was collected from a three-phase olive mill. The inoculum used was collected from four, six-liter digesters operating under steady state conditions with cattle manure as substrate and an OLR of approximately 3kg/VSm3-d. Protein-rich substrates were collected from a pet food manufacturer. All substrates were stored in a freezer at -20°C after collection until further utilization.

In this work 250mL Duran bottles were used as batch digesters and for assessing the biomethane potential and biodegradability of the different substrates. In order to evaluate the batch process and generate comparable results two loading rates (10 and 20 $kgVS/m^3$) were evaluated. The I/S ratio was set at 2 and 1 respectively. In the second stage of the experiment, four identical stainless steel reactors were used. The reactors were heated through the recirculation of hot water within the water jacket, feeding and removal of the substrates was taking place manually, while the mixing of the reactors was achieved with the application of immersed propellers and direct current mixing motors.

Results and discussion

Mono digestion of the substrates

The highest biogas production of the substrates under mono-digestion conditions was obtained using bone and meat meal with a value of 511 $mLCH_4/gVS_{added}$. This was approximately 16% higher compared to the production of biogas using blood meal and feather meal mixture with a production of 427 $mLCH_4/gVS_{added}$. Related to the plant-derived substrates the specific methane production of gluten and soy bean meal amounted to 495 $mLCH_4/gVS_{added}$ and 460$mLCH_4/gVS_{added}$ respectively.

Co-digestion of OME with high nitrogen substrates

In the second stage of this study OME was co-digested with high nitrogen substrates. The highest biogas production was attained using a combination of OME and fishmeal with a yield of 482 $mLCH4/gVS_{added}$ corresponding to the 77% of the ultimate production for this mixture. Digestion of the other mixtures gave also very positive results with the mixture of OME & blood meal and of OME & soybean meal offering yields of 454 and 446 $mLCH4/gVS_{added}$ corresponding to the 74 and 72% of the ultimate production for the two mixtures respectively. The lowest biogas production registered was the one obtained from a mixture containing OME & bone and meat meal with a yield of 418 $mLCH4/gVS_{added}$, equal to the 68% of the theoretical value.

Validation of the results in CSTR systems

In order to validate the results of the previous experimental stages, the biogas yield using four substrate mixtures composed of olive mill wastewater (OMWW) and protein-rich substrates were assessed in semi-continuous digesters. Feeding was provided daily throughout the week including the weekends. The four substrates chosen for analysis in the semi continuous systems were:

(a) bone and meat meal as it is the co-substrate with the highest market availability and the lowest price.

(b) blood meal; this substrate was selected due to its negative effects recorded when high loading rate was used in the batch digestion.

(c) fishmeal; this substrate offered the highest bio-methane production of all substrate mixtures assessed in the batch digestion stage.

(d) The last of the substrates assessed was gluten. This was the only plant-derived protein-rich substrate assessed; its selection was based on its higher protein content over the second candidate substrate, namely soybean meal.

In this experimentation stage the highest biogas production was achieved by the mixture containing blood meal with an average production equal to 449 mLCH4/gVS added, followed by biogas production from bone and meat meal with 443 mLCH4/gVS added. The lowest production was achieved when using the mixture containing FM with an average yield of 416 mLCH4/gVS added. Our results indicate that the type of protein added is insignificant; in all cases the yields of the four different mixtures were within 7% of each other ranging between 449 and 416 mLCH4/gVS added. In all cases VFA concentrations were very low and no inhibition was recorded in any of the reactors treating the different substrates.

Conclusion

OMWW, a heavily polluted effluent can be valorized through the anaerobic digestion process into biogas and digestate. The addition of nitrogen-rich substrates offers the opportunity to achieve a steady process with high bioconversion efficiency. Furthermore, the type of the co-substrate seems to be insignificant as no substantial differences in biogas yield were registered when fish, animal or plant derived proteins were used as nitrogen-carrying additives. This work offers a framework for the possibility to valorize the OMWW in treatment plants in the vicinity of the olive mills without the requirement of manure addition as carrier substrate. This development would minimize the risk of bacterial contamination from transportation of manure near olive mills. According to our results one ton of nitrogen-rich substrates can be used for the treatment of more than $50m^3$ of OMWW while at the end of the process the recovered effluent has significantly better characteristics as soil improver when compared to raw OMWW.

492445 Evaluation and Comparison ofÂ the Scale-up Potential for Autotropic and Heterotrophic Algal Oil Production

Abstract:

Significant research has been completed over the past decade to develop technologies to facilitate the production of algal lipids for use as feedstock for next generation renewable fuel facilities. The interest for microalgae is motivated by the substantial increase in yield per land area compared to oilseed crops reducing the pressure to supplant food crops with energy crops. However, the challenges to produce oil from algae at scales approaching crop oil production facilities have been great. One of those barriers is the need to get photons from sunlight or artificial sources to the growing algae as an energy source. It has recently been proposed that growing the algae heterotropically might overcome this scale-up barrier. The transition to heterotrophic halts the photosynthesis process, and requires an organic carbon source to provide energy, as the heterotrophic strain of microalgae is not able to produce energy compared to the autotrophic which produces an energy source though photosynthesis. Additionally, the transition has been proven to increase the lipid content of the microalgae by replacing the chlorophyll cells produced during photosynthesis. If the strain of microalgae used is non-light dependent it will negate the requirement of industrial scale, clear photo bioreactors or open ponds. Additional chemicals and catalysts will be required for the growth and development of the microalgae strains for the growth media.

In this study, a group of senior chemical engineering students developed preliminary designs and then performed an economic analysis for algal oil production utilizing both autotropic and heterotrophic growth strategies. Both proposed systems utilizes the strain of microalgae *Chlorella Vulgaris* as the raw material. *Chlorella Vulgaris* has a lipid content ranging between 15-35 wt%.

The study involved the development of Input/output diagrams, block flow diagrams, and process flow diagrams for compete processes starting from lab algae inoculation through to final lipid oil product purification. Preliminary equipment sizing was performed. The capital costs were estimated using a factored broad cost estimation method based on vendor quotes for the largest equipment and cost chart data for the rest of the equipment. Revenues and operating costs for a 20 year project life were also estimated.

The heterotrophic option will yield a higher lipid content which will generate a larger amount of oil when compared to the autotrophic strain but will require a higher quantity of inputs. The production throughput was comparable to a small scale oilseed production facility. Neither option is currently economically competitive with a commercial soybean oil processing facility. However, the heterotrophic option had significant advantages over the autotropic design. The key barriers to commercialization were identified so that future R&D activities can focus in these areas.

MOMENTIVE
inventing possibilities

Process Developments in Continuous Silicone-Polyether Copolymers

- Akshay Kundan
- Chauncey Rinard
- Steven Bahr

10/30/2017

BACKGROUND ON SILICONE-POLYETHER COPOLYMERS (SPC)

MOMENTIVE

Urethane Additives

POTENTIAL APPLICATIONS, KEY PRODUCTS & TYPICAL BENEFITS

- Improve reaction mix compatibility
- Determine reaction rate and productivity
- Provide bulk foam stabilization
- Regulate cells size and openness
- Make smooth surface of molded foam
- Improve properties like fire, comfort
- Improve dimensional stability of low-density PU
- Help CASE systems for easy pouring and faster cure
- Make PU leather smooth and abrasion-resistant
- Enhance foam physical properties and fire behavior

MOLDED FOAMS

- Automotive seating
- Furniture
- Pillows and mattresses

SLAB FOAMS

- Furniture
- Bedding
- Technical foams
- Laminated foams

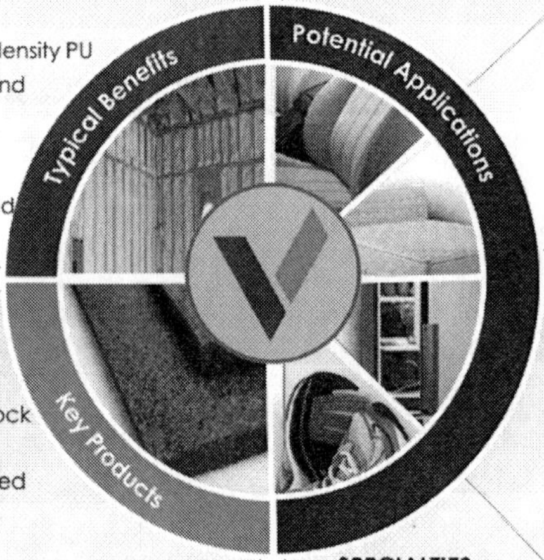

- Niax* silicones and catalysts for slabstock foams (conventional, high resilience and visco-elastic)
- Geolite* process modifiers for slabstock foams
- Niax silicones and catalysts for molded foams based on TDI, MDI and TM systems
- Niax silicones, catalyst and special additives for rigid foams
- Niax additives for specialty PU applications
- Fomrez[†] special PU catalysts

RIGID FOAMS

- Fridge, cool store, water heater, display units
- Refrigerated trucks, reefer containers
- Construction panels, SIPs
- Insulation panels, boards
- Gap filling
- Insulated pipe & tanks
- Mining/grouting
- Decorative, structural, packaging, floral

SPECIALTIES

- Carpet backing
- Electronics
- Footwear
- Artificial turf
- Artificial leather

3

MOMENTIVE

*Niax and Geolite are trademarks of Momentive Performance Materials Inc.
[†]Fomrez is a trademark of Chemtura Corporation, used with permission by Galata Chemicals LLC.

Momentive Urethane Additives Portfolio

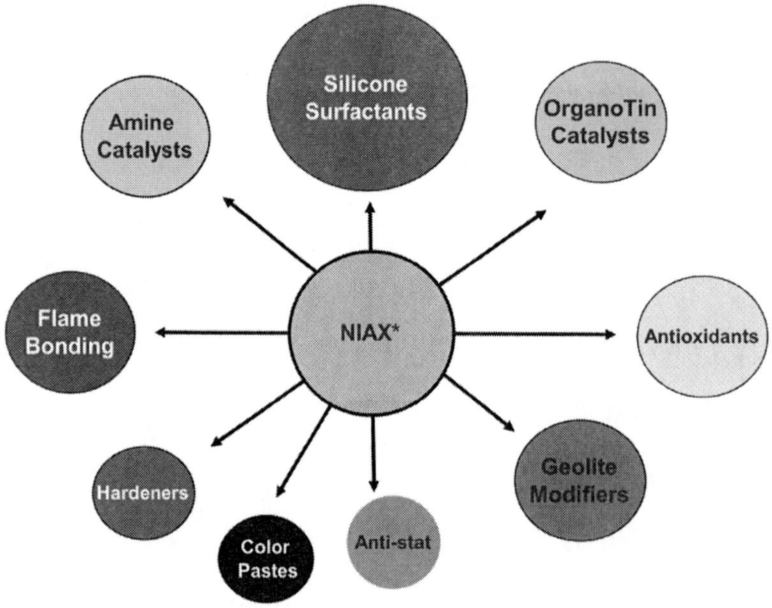

4

MOMENTIVE

Why Silicone Surfactants?
Emulsification and Stabilization

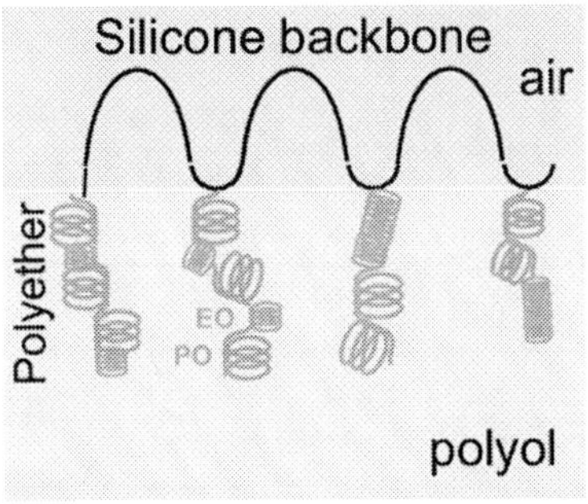

Surfactant 1% Surfactant in a 3000 MW Triol	γ (dynes/cm)
None (pure polyol)	**35**
Nonylphenolethoxylate	**32**
Sulfonated Fatty Ester	**31**
Silicone/Polyether	**25***
Fluorocarbon/Polyether	**24**

*The only product that will stabilize a urethane foam. Surface Tension reduction is a necessary but not sufficient requirement for foam stabilization

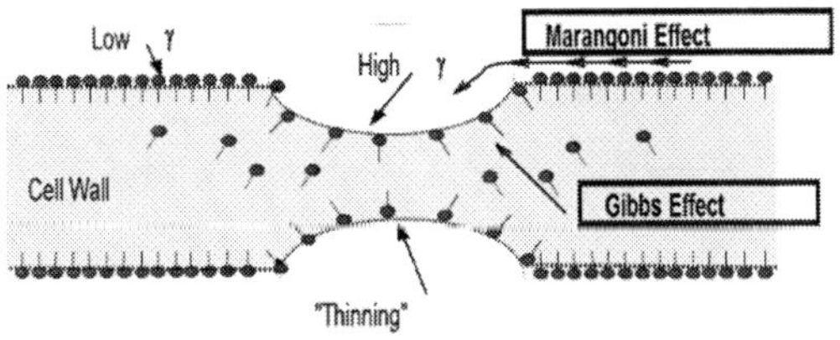

Gibbs-Marangoni: Thinning causes lower local concentration of surfactant which raises γ causing a γ gradient in the film. This gradient is a restorative force drawing more surfactant to lower the γ. The surface is said to become "elastic".

Ewers-Sutherland: Similar concept except this approach suggests that as the surfactant mono-layer moves to reduce the local γ gradient it drags with it sufficient bulk material to help heal the rupture.

5 MOMENTIVE

Copolymer Chemistry

Siloxane backbone

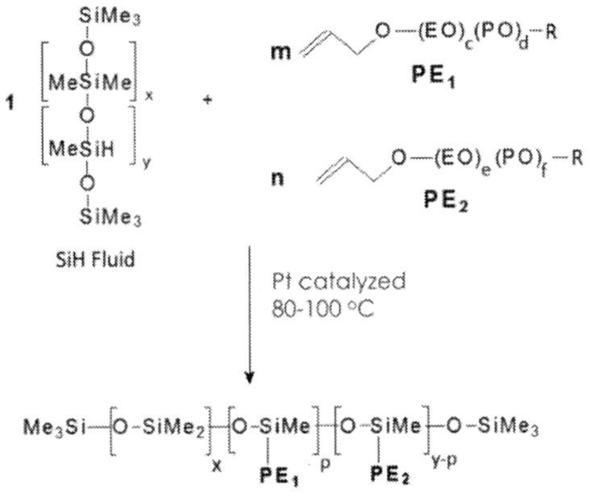

Polyethers Pendants

Silicone Polyether Copolymer

- The reaction is governed by compatibility, reaction selectivity and randomization.
- Potential Technologies in MPM:
  - Batch process
  - Continuous Copolymer Unit (CCU)
  - Static mixing plug flow reactors
  - Multistage Mixer (MSM)
  - Patented Technologies: CCU, Static mixing plug flow reactors, MSM.

6 MOMENTIVE

IP REVIEW

MOMENTIVE

Innovation – Continuous SPC Manufacturing Technology

- Detailed IP search of available patents for producing silicone copolymers continuously.
- Process development boom came in early 2000s.

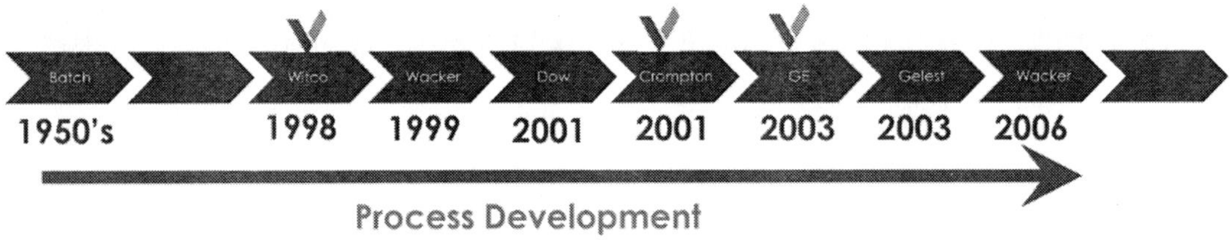

Year filed	Company	Patent #	Process
1998	**Witco Corporation**	**US 5986022**	**CSTR and PFR (CCU)**
1999	Wacker Chemie Gmbh	US6350824 B1	Pre-mixed and fed into the reaction zone
2001	**Dow Corning Toray Silicone**	**US 6,410,772 B2**	**CSTR, tubular reactor and PFR**
2001	Crompton	US20020099137 A1	Static Mixing PFR in series or parallel
2003	**General Electric Company**	**US 6,897,280 B2**	**Multi-stage blade – mixed plug flow reactor (MSM)**
2003	Gelest Inc	US 20040176561 A1	PFR and a tubular reactor packed with a solid, acidic polymerization catalyst
2005	Wacker Chemie Ag	US7145028 B2	Integrated loop tube reactor

8 MOMENTIVE

CONTINUOUS COPOLYMER UNIT (CCU)

PATENT NO.: 5986022
YEAR APPROVED: 1999
ASSIGNEE: WITCO CORPORATION

MOMENTIVE™

CCU Design

- 2 CSTR's and 1 PFR in series
- Reaction has to complete at least 80-85% before entering PFR to avoid phase separation.
- Properties slightly different from batch products.
- CSTR Limitation: Not all of any of the reactants can be consumed completely.

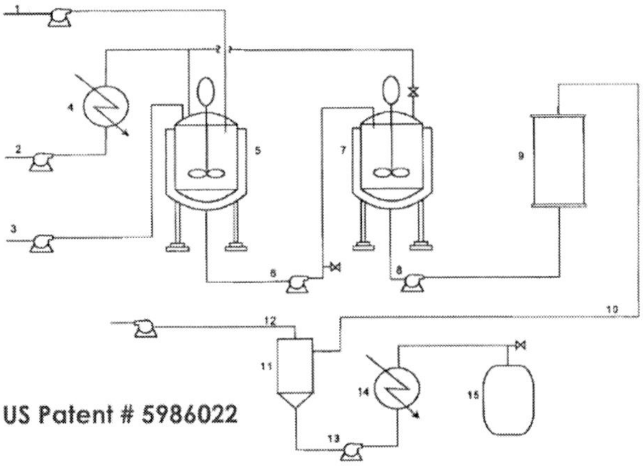

US Patent # 5986022

- PFR Limitation: Immiscible SiH fluid and olefinic compound will phase separate very rapidly and reaction slows down or ceases completely.
- Benefits from a batch: Low Operating Cost (Conversion Cost Equivalent to LCC), Reduced Footprint, Reduced Waste, Solvent Reduction, Capacity.
- Cons: Versatility

Has its advantages over batch but versatility is missing.

MOMENTIVE™

MULTI-STAGE MIXER PATENT (MSM)

PATENT NO.: US 6,897,280 B2
YEAR APPROVED: 2005
ASSIGNEE: GENERAL ELECTRIC COMPANY

MOMENTIVE

MSM Patent Overview

- Evolution of CCU Technology.
- ID 50 mm, L – 0.6 m, 8 stages of 4-8 min residence time, tubular jacketed reactor. Offset channels to prevent back mixing.
- Eddies and vortices of sufficient intensity in multi-stage blades shear biphasic liquid mixture and disperse one phase into another for rapid reaction.
- Last stage of reactor is plug flow. 75-90% of reaction is finished before entering this region.
- Front end equipment required
  Flow meters, buffer, catalyst addition systems, stripping unit
- Back end equipment required
  Product stripper, product cooler, product filters and blend equipment (tanks or mixer)

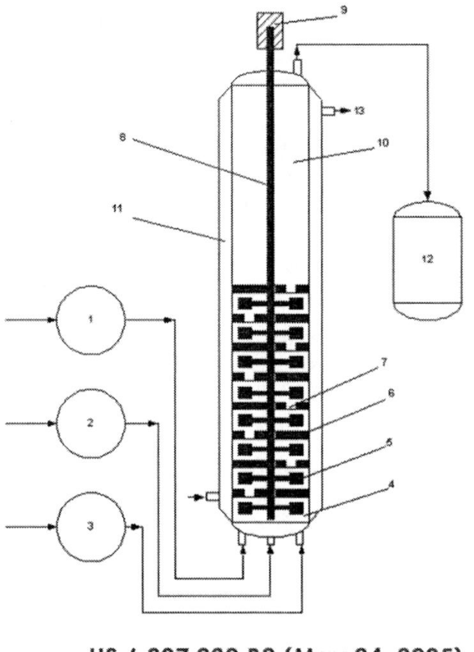

US 6,897,280 B2 (May 24, 2005)

2nd Generation CCU

MOMENTIVE

Batch/CCU/MSM

Comparison of processes for different SPC (US Patent# US 6,897,280)

Raw material	Process	SiH (cc H2/g)	Appearance	Viscosity (cst)	Foam Rise (cm)	Flow Air (porosity)	Cell Structure
Siloxane A + PE_1 + PE_2 + Buffer + Pt catalyzed	Batch	ND	Clear	2349	38.0	5.2	7
	CCU	**0.4**	**Hazy**	2461	**36.8**	4.2	**10**
	MSM	ND	Clear	2482	38.6	6.3	7
Siloxane B + PE_3 + PE_4+ Buffer + Pt catalyzed	Batch	ND	Clear	2771	40.5	2.2	6
	CCU	**0.1**	**Hazy**	2569	**Total Foam Collapse**		
	MSM	ND	Clear	2626	39.2	7.0	7
Siloxane B + PE_5 + PE_4 + Buffer + Pt catalyzed + DPG	Batch	ND	Clear	1003	37.9	6.6	7
	CCU	ND	Clear	942	37.1	6.1	**8**
	MSM	ND	Clear	1036	37.9	5.6	7
Siloxane B + PE_6 + PE_4 + Buffer + Pt catalyzed	Batch	ND	Clear	1794	40.9	5.1	7
	CCU	ND	Clear	1851	39.0	5.9	7
	MSM	ND	Clear	1833	40.7	5.1	7

MSM produced similar to batch grade. CCU is slightly different.

13

MOMENTIVE™

DOW PATENT

PATENT NO.: US 6,410,772 B2

YEAR APPROVED: 2002

ASSIGNEE: DOW CORNING TORAY SILICONE CO., LTD.

MOMENTIVE™

Dow Patent Overview

- ID: 50 mm, L = 1 m, 33 disks
- Blade or pin stirrers severely impact plug flow conditions.
- 2 or more disks maintain plug flow conditions.
- Rotation of disks facilitates mixing of various components & improves efficiency while maintaining plug flow conditions.
- Reactor capacity significantly improved by adding a pre-mixer.
- Static reactor downstream (25 mm ID, L - 3m, 1.5L) solves the feed rate & elongation of tubular reactor issues.
- Loading of SiH fluid to PE is about 1.
- Residence time of 4 min for feed rate of 50kg/hr achieved.
- **Limitations:** Single PE product.

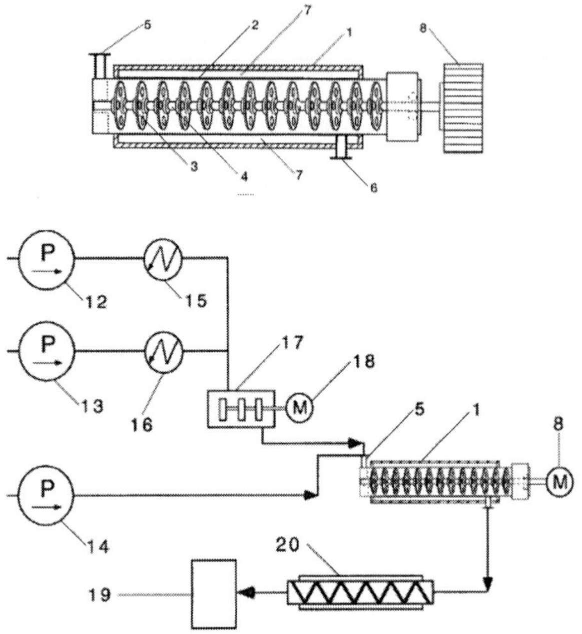

US Patent # 6,410,772 B2

MOMENTIVE

Summary of Comparisons

	Batch	CCU	MSM
Reactor	Single tank	At Least One 100-300 Gallon Tank Plug Flow Reactor	Tubular reactor with multiple stages of stirring blades
Pros/Cons	Versatility	Low Operating Cost (Conversion Cost Equivalent to LCC) Reduced Footprint Reduced Waste Solvent Reduction Capacity	Low Operating Cost (Conversion Cost Lower than LCC) Versatility Small Footprint Reduced Waste Solvent Reduction Capacity
Capability	Terminal/Pendant Copolymers Single/Multiple Polyethers Wide Range of MW's/Viscosities	Terminal/Pendant Copolymers Single Polyether Lower End MW's/ Viscosities	Terminal/Pendant Copolymers Single/Multiple Polyethers Wide Range of MW's/Viscosities
Product grade		Different than batch	Similar to batch

Process Development key to achieving good quality SPC

MOMENTIVE

Thank You.

DISCLAIMER: THE MATERIALS, PRODUCTS AND SERVICES OF Momentive Performance Materials Inc. and its subsidiaries and affiliates DOING BUSINESS IN LOCAL JURISDICTIONS (collectively "SUPPLIERS") ARE SOLD BY THE RESPECTIVE LEGAL ENTITY OF THE SUPPLIER SUBJECT TO SUPPLIERS' STANDARD CONDITIONS OF SALE, WHICH ARE INCLUDED IN THE APPLICABLE DISTRIBUTOR OR OTHER SALES AGREEMENT, PRINTED ON THE BACK OF ORDER ACKNOWLEDGMENTS AND INVOICES, AND AVAILABLE UPON REQUEST. ALTHOUGH ANY INFORMATION, RECOMMENDATIONS, OR ADVICE CONTAINED HEREIN IS GIVEN IN GOOD FAITH, SUPPLIERS MAKE NO WARRANTY OR GUARANTEE, EXPRESS OR IMPLIED, (i) THAT THE RESULTS DESCRIBED HEREIN WILL BE OBTAINED UNDER END-USE CONDITIONS, OR (ii) AS TO THE EFFECTIVENESS OR SAFETY OF ANY DESIGN INCORPORATING SUPPLIERS' PRODUCTS, MATERIALS, SERVICES, RECOMMENDATIONS OR ADVICE. AFOREMENTIONED EXCLUSIONS OR LIMITATION OF LIABILITY ARE NOT APPLICABLE TO THE EXTENT THAT THE END-USE CONDITIONS AND/OR INCORPORATION CONDITIONS CORRESPOND TO THE RECOMMENDED CONDITIONS OF USE AND/OR OF INCORPORATION AS DESCRIBED BY SUPPLIER IN ITS PRODUCT DATA SHEET AND/OR PRODUCT SPECIFICATIONS. EXCEPT AS PROVIDED IN SUPPLIERS' STANDARD CONDITIONS OF SALE, SUPPLIERS AND THEIR REPRESENTATIVES SHALL IN NO EVENT BE RESPONSIBLE FOR ANY LOSS RESULTING FROM ANY USE OF ITS MATERIALS, PRODUCTS OR SERVICES DESCRIBED HEREIN. Each user bears full responsibility for making its own determination as to the suitability of Suppliers' materials, services, recommendations, or advice for its own particular use. Each user must identify and perform all tests and analyses necessary to assure that its finished parts incorporating Suppliers' products, materials, or services will be safe and suitable for use under end-use conditions. Nothing in this or any other document, nor any oral recommendation or advice, shall be deemed to alter, vary, supersede, or waive any provision of Suppliers' Standard Conditions of Sale or this Disclaimer, unless any such modification is specifically agreed to in a writing signed by Suppliers. No statement contained herein concerning a possible or suggested use of any material, product, service or design is intended, or should be construed, to grant any license under any patent or other intellectual property right of Suppliers or any of its subsidiaries or affiliates covering such use or design, or as a recommendation for the use of such material, product, service or design in the infringement of any patent or other intellectual property right.

Momentive and the Momentive logo are trademarks of Momentive Performance Materials Inc.

17 MOMENTIVE

497659 Esterification of Carboxylic Acids Present in the BIO-OIL Produced By Thermal Cracking of Triacylglycerols

Abstract:

The production of alternative fuels from renewable sources has been object of several studies in recent years. A promising alternative is the conversion of biomass to biofuels by thermal cracking. Through this process, it is possible to obtain a liquid product called bio-oil, with physicochemical properties similar to the petroleum crude oil. However, bio-oil might contain a high amount of carboxylic acids that results on a high acid index ranging from 110 mg KOH/g of sample to 207 mg KOH/g of sample, whereas the maximum required by the current legislation (EN 14104) is 0.5 mg KOH/g of sample. In this work, the bio-oil was produced from commercial soybean oil at 525°C and a mass flow rate of 350 g/h. The objective of this study is to perform esterification reactions in order to reduce the acidity index of the bio-oil. These reactions present a methodology of easy operation, low cost and may reduce the bio-oil corrosivity. For this, alcoholic reagents, ethanol and methanol, were used for the esterification reactions. The influence of different temperatures, reaction times, catalysts and, stoichiometric ratios between the reagents on the reduction of the acid number and improvement of the bio-oil quality were evaluated. Good results were obtained with the esterification reaction using ethanol, sulfuric acid, bio-oil to ethanol ratio of 10:5, temperature of 60°C and reaction time of 90 minutes. The acid index of the bio-oil sample decreased from 111.33 mg KOH/g of sample to 34.64 mg KOH/g of sample, resulting in a reduction of approximately 70%.

500673 Chiral Control of Enantiomers through Crystallization: Moving from Ternary Phase Diagrams to Design Spaces

Abstract:

Although chiral purity is a critical quality attribute for the majority of synthetic drugs being developed, the understanding of process design for chiral purity in terms of conventional batch crystallization processes is limited. The industrial implementation within this area has relied heavily on the construction of ternary phase diagrams, system classification, and the use of subject matter experts to determine thermodynamic operating policies. These operating policies have been developed to yield resulting solids with 100% ee; regardless of the product specification/target and the potential upgrades to ee remaining between the molecule being isolated and the drug substance. In the present work (which for simplicity is limited to the case of a single chiral center in the absence of salts, polymorphs, and solvates/hydrates), a new policy is proposed for obtaining resulting solids with ee < 100% but passing specifications for arbitrary systems. This operating policy can provide significant isolation yield improvements when compared to the conventional policy. Additionally, this work demonstrates an approach for the creation of design spaces for arbitrary ternary systems based on the collection of limited small (<10 g total) scale solubility data. This approach has been extended to processes with sequential re-crystallizations and can be used to determine â€œglobalâ€ optimal policies for an entire synthetic route.

499039 Leveraging Mechanistic Understanding of Impurity Rejection during Crystallization for Rational Process Design

Abstract:

The control of impurities in the chemical synthesis of a new drug substance is a requirement in registering pharmaceuticals with worldwide regulatory agencies. Crystallization is an essential tool in controlling the purity profile of a drug substance and its intermediates. An overview of the various mechanisms of impurity incorporation during crystallization will be discussed along with remediation strategies. A case study will be provided on the replacement of chromatography with crystallization during the development of a second generation commercial manufacturing route for Ceftolozane, the cephalosporin antibiotic found in Zerbaxa. The implementation of a crystallization based purification process for Ceftolozane, replacing a chromatography/nanofiltration/diafiltration sequence, resulted in a more productive, greener, and more portable manufacturing process. Another case study will be presented on limiting substrate concentration on product crystals during a reactive crystallization via engineering design.

488309 Development of Combined Continuous Reaction to Crystallization Process Using PAT Tools

Abstract:

Introduction

Figure 1. Synthesis of (*Z*)-3-chloro-2-(phenylthio)-*N*-(p-tolyl)acrylamide

Traditionally the majority of active pharmaceutical compounds have been manufactured at large plant scales using batch methodologies. Continuous manufacturing is becoming an increasing attractive alternative due to apparent benefits such as smaller manufacturing footprints, increased process safety, lower manufacturing costs and increased supply-chain flexibility. In this work, a new approach to the design of continuous manufacturing platforms is being presented which is not restricted to using such conventional equipment but which is based around bespoke design and manufacture. As a model compound, 2-chloro-*N*-(4-methylphenyl)propanamide (**S1**), a precursor in the multi-step synthesis of (*Z*)-3-chloro-2-(phenylthio)-*N*-(p-tolyl)acrylamide (**P1**, Figure 1) is used. The aim of this work is to design a combined continuous reaction to crystallization process for **S1** and overcome the existing challenges between the two process.

Solubility Studies for S1

Figure 2. Solubility data of **S1** in toluene + hexane mixtures: *, $w_2 = 1$; +, $w_2 = 0.8$; ◊, $w_2 = 0.6$; □, $w_2 = 0.4$; -, Phase Equilibria with NRTL equation calculated. w_2 is the mass fraction of toluene in binary toluene + hexane mixtures.

The solubility behaviour of **S1** in ethyl acetate + hexane, toluene + hexane, acetone + hexane and butanone + hexane solutions were measured as the solubility characteristics have considerable influence upon the design and optimisation of any crystallisation process, affecting both the resulting crystal structure and crystal size distribution (CSD)[1]. The solubility study is done using the polythermal method and conducted in EasyMax (Mettler Toledo) reactor with in situ Focused Beam Reflectance Measurement (FBRM G400, Mettler Toledo) to characterize the saturation temperature of **S1**. The experimental data are correlated with Apelbat, λh and the Phase Equilibria with NRTL (non-random two liquid) model equations using the function *lsqcurvefit* in Matlab's optimization toolbox to enable both interpolation and extrapolation of the measured data, as shown in Figure 2.

Design of Continuous Reaction and Crystallization Process

Reaction: Using a modified procedure developed by Li *et al*[2], a batch process is employed where a-chloropropionyl chloride (12 mmol) was added dropwise to a suspension of *p*-toluidine (10 mmol) in toluene (20 mL) at 0°C and the resulting solution was heated to reflux for 1 h with vigorous stirring. After cooling, the solvent was removed under vacuum and the resulting off white solid was collected by filtration and washed thoroughly with cold cyclohexane. Spectral data of the obtained product matched that reported in the literature[3]. Therefore, it has been decided to use toluene as the reaction solvent as it can also be used as a crystallization solvent based on the solubility studies. However, this developed procedure produces a small amount of HCl as a by-product, which can be problematic for reaction product purity and its subsequent crystalisation steps. Therefore, it is decided to run the reaction in a novel 3D printed continuous biphasic reaction system in the presence of aqueous sodium hydroxide to neutralize HCl salt. The design of the biphasic reactor is based on the specific needs of the reaction which is capable of achieving extremely high dispersion rates between the two phases under the very low Reynolds numbers typically encountered in the pharmaceutical industry ($Re < 5$). Process steps are quantified in detail using methods such as computational fluid dynamics (CFD) leading to tailored equipment designs which can be manufactured with high precision using metallic 3D printing. The biphasic reactor is coupled with liquid-liquid separation such that both unit operations are designed using fundamental engineering principles combined with numerical analyses (based upon the requirements of the specific process), prior to being 3D printed and combined to create a bespoke, sealed, hard-piped manufacturing system.

Reaction to Crystallization: From the liquid-liquid separator, the organic phase is transferred in the feed/dissolution tank while aqueous phase goes to waste. In situ FTIR (React IR 15, Mettler Toledo) is used on the feed/dissolution tank to determine the concentration of **S1** prior to crystallization and the crystallization kinetic data obtained (explained below) enables the adjustment of crystallization parameters for better control and set up of the process.

Crystallization: Using the solubility curve obtained from solubility studies for toluene, a continuously operated single stage mixed suspension, mixed product removal (MSMPR) crystallizer with intermittent withdrawal via dip pipe using an automated pressure supply[4-6] is set up for the cooling crystallization of **S1**. Acquisition of reliable kinetic correlations that describe how rates of nucleation and growth vary over the design space for the crystallization process is necessary for enabling robust prediction of crystallizer performance. Therefore, simple extraction of the growth and nucleation parameters from the single stage MSMPR operating at steady state is performed using the population balance model. Comprehensive experimental data for model fitting and parameter estimation is obtained by varying crystallizer operating conditions (temperature, supersaturation and suspension density) to induce changes in the rate-affecting variables for growth and nucleation. FBRM is used for real-time monitoring of size distribution and identification of steady state and hence enabling on-demand adjustment of operating conditions to transition multiple steady-states for rapid acquisition of kinetic data while PVM (ParticleView V819, Mettler Toledo) provides microscopy quality images in real time to monitor the crystal habit. Malvern Mastersizer is used as an offline analysis to determine the crystal size distribution (CSD) at steady-state while the concentration of mother liquor at steady-state is determined gravimetrically. By estimating the six kinetic parameters for crystal growth and nucleation, accurate prediction of CSD and concentration results at steady-state can be achieved.

References:

[1] Mullin, J. W. *Crystallization*; Butterworth-Heinemann, 1997

[2] Li, X.; Wang, D.; Wu, J.; Xu, W. Novel Approach to 3-Methyl-1H-quinoxalin-2-ones. *Synthetic communications* **2005,** *35* (19), 2553.

[3] Murphy, M.; Lynch, D.; Schaeffer, M.; Kissane, M.; Chopra, J.; O'Brien, E.; Ford, A.; Ferguson, G.; Maguire, A. R. Investigation of the synthetic and mechanistic aspects of the highly stereoselective transformation of [small alpha]-thioamides to [small alpha]-thio-[small beta]-chloroacrylamides. *Organic & Biomolecular Chemistry* **2007,** *5* (8), 1228.

[4] Hou, G.; Power, G.; Barrett, M.; Glennon, B.; Morris, G.; Zhao, Y. Development and Characterization of a Single Stage Mixed-Suspension, Mixed-Product-Removal Crystallization Process with a Novel Transfer Unit. *Crystal Growth & Design* **2014,** *14* (4), 1782.

[5] Power, G.; Hou, G.; Kamaraju, V. K.; Morris, G.; Zhao, Y.; Glennon, B. Design and optimization of a multistage continuous cooling mixed suspension, mixed product removal crystallizer. *Chemical Engineering Science* **2015,** *133*, 125.

[6] Morris, G.; Power, G.; Ferguson, S.; Barrett, M.; Hou, G.; Glennon, B. Estimation of Nucleation and Growth Kinetics of Benzoic Acid by Population Balance Modeling of a Continuous Cooling Mixed Suspension, Mixed Product Removal Crystallizer. *Organic Process Research & Development* **2015,** *19* (12), 1891.

485308 Process Design to Preserve Particle Properties during Crystallization of Energetic Material

Abstract:

Obtaining desired particle size and morphology of isolated material from a quenched crystallization can be challenging during scale up. This work focuses on the scale-up of a quenched crystallization of an energetic compound while maintaining a large blocky morphology of the isolated material. The compound has low solubility in most solvents, therefore recrystallization options were limited. In this case, the key to obtaining the desired particle size and morphology was quenching the reaction mixture into a nitrate salt solution. This talk will discuss the process optimization and the effects of mixing, temperature, and salt concentration on this quenched crystallization. Optimization and development efforts were performed in the 1-L Mettler Toledo OptiMax reactor before being scaled up to the pilot plant.

491816 Multi-Scale and Multi-Purpose Modelling for *in silico* Designing and Optimization of Pharmaceutical Crystallization Process

Abstract:

Crystallization often serves as a key final and intermediate separation/purification unit operation in pharmaceutical production. More than 90% of small molecule drugs are produced in crystalline form[1]. Operation conditions of crystallization process determine process metrics such as crystal size distribution, yield, space-time yield (STY) and crystal quality attributes like crystal shape, morphology as well as solvates. The crystal product properties affect the efficiency of downstream unit operations such as filtration, drying and formulation. Moreover, in pharmaceutical products the crystal properties affect the stability and efficacy of drugs. Therefore, a comprehensive multi-scale and multi-purpose thermodynamic and kinetic modeling is the main objective of this work in order to understand crystallization mechanism and define optimum operation conditions. Optimization of crystallization operation conditions includes selection of the best solvent or mixture of solvents, cooling rate, anti-solvent flowrate, pH profile, mixing and seeding parameters as well as selection of mode of operation such as batch or continuous. Figure 1 shows a comprehensive multi-scale and multi-purpose modeling framework for *in silico* designing and optimization of pharmaceutical crystallization process.

Figure 1: A schematic diagram illustrating comprehensive multi-scale and multi-purpose modeling framework for *in silico* designing and optimization of pharmaceutical crystallization process

The framework will be implemented in Matlab-Simulink environment as a tool to support different engineering tasks from solvent selection to crystallization process design, optimization and control in life science and biopharmaceutical industries.

Acknowledgment: We would like to thank the Danish Council for Independent Research (DFF) for financing the project with grant ID: DFF-6111600077B.

---

[1] Narayan Variankaval, Aaron S. Cote, *AIChE* 54 (2008) 1682 -1688

492269 Sensitivity Analysis and Optimization Study of Pharmaceutical Crystallization Process

Abstract:

Crystallization is a prevalently applied technique in many industries including food, bulk and fine chemicals as well as manufacturing of active pharmaceutical ingredients (APIs). In the pharmaceutical industry, crystallization processes are employed mainly for separation,, purification as well as for formulation of APIs. Moreover, the process should guarantee the performance of the delivered APIs suitable for downstream processing, handling and drug efficacy.

The performance of an API crystal is strongly affected by the product quality attributes such as crystal size distribution, morphology, shape and purity, which are the main targets of the process output to be achieved within specific range. However, in a crystallization process, complex feedback interactions take place between the formed crystals and the growth and nucleation kinetics. These are at the same time dependent on the supersaturation and hydrodynamic conditions to which the crystals are subjected. Consequently, the quality characteristics of the crystal product have a complex dependency on multiple operational variables and input parameters.

In order to evaluate and quantify which process factors or variables have the most impact on the final quality attributes of the crystal product, a comprehensive sensitivity analysis of the system model is required. This study therefore focuses on a global sensitivity analysis of a predictive pharmaceutical crystallization model through assessing the sources of uncertainty and their influence on crystal size distribution as the measured process outcome/performance criteria in order to guide design and suggest the optimized operation conditions.

This process model of pharmaceutical batch cooling crystallization is developed based on compartmentalization technique which takes into account mixing phenomena as well. Several global sensitivity analysis methods including Morris screening, global derivative based sensitivity analysis (DGSM), variance decomposition methods such as Sobol's index and linear regression of Monte Carlo outputs are used to perform the sensitivity analysis. The developed crystallization process model is analyzed for a range of values for parameters like cooling rate, growth and nucleation kinetics etc. to rank their contributions on the measured outcome of the system "crystal size distribution". Comparison of two sensitivity analysis methods and evaluation of the results are used as a guide for process optimization and robust system design.

Acknowledgment

We would like to thank the Danish Council for Independent Research (DFF) for financing the project with grant ID: DFF-6111600077B.

495572 Segmented Continuous Crystallization Process Optimization

Abstract:

The continuous crystallization behavior of glycine in segmented tubular crystallizer has been investigated. The factors of liquid slug length and shape, Reynolds number, suspension density and arrangement were investigated respectively on the crystal growth rate, size distribution and morphology. The results found that the slug being closer to a sphere both in length and shape is the best compared to the other slug geometries, in which mixing within each stable self-circulating slug is maximized, thus the crystal growth rate is faster and the particle size distribution is more uniform. Besides, Higher Reynolds number improves the growth rate and uniform size distribution at the same residence time. The mass fraction of seed crystals at 2% was also determined to be the optimal condition. The arrangement of the tube in vertical and horizontal direction has little influence on the final yield and crystal characteristics. These results provide an insight into the segmented tubular continuous crystallization.

499049 Biochar Collection Overview and Design Upgrades in Biomass Pyrolysis Pilot Plant

Abstract:

The Thermochemical Process Development Unit (TCPDU) at the National Renewable Energy Laboratory conducts research on a pilot plant that may be configured for biomass fast pyrolysis or gasification with the ultimate objective of creating renewable transportation fuels. Current research is focused on optimizing pyrolysis conditions for low-cost feedstock blends that are widely available and relevant on an industrial scale. Future research (2018) will focus on ex-situ catalytic upgrading of pyrolysis vapors using a riser reactor and regenerator system, similar to a fluid catalytic cracking system. The TCPDU is capable of feeding ½ ton of biomass per day and commonly operates at a feed rate of ~33 lb/h biomass while configured for fast pyrolysis. At this feed rate, the TCPDU generates ~2.2 gal/h of raw bio-oil and ~4.5lb/h of solid bio-char.

Efficient solids separation and collection is essential when thermochemically converting biomass to desirable products. Solids in the TCPDU are removed from the process stream via dual cyclonic separators. When solids are not collected efficiently, carryover to downstream operations occurs often leading to process disruption, requiring downtime and maintenance before operations resume. Whenever operators must perform maintenance that involves opening the process they are subjected to additional exposure hazards. This increased downtime is very expensive, and therefore must be minimized for a plant to operate efficiently.

While running experimental feedstocks within the TCPDU, the solids collection system was quickly determined to be insufficient for the nature of the solids generated when varying feedstocks and run parameters. In order to conduct our research, it was deemed necessary to redesign our solids collection systems. A 3-phase approach was taken in an effort to continue performing research while improving our solids collection system in a step wise manner. Phase 1 included an assessment of the system as configured with a focus on minor changes that could improve operations in the short term. Phase 2 consisted of a thorough assessment of our solids collection system in an effort to identify underlying causes of solids collection problems, with a focus on design changes that would improve flexibility across feedstocks and run conditions. Phase 3 involved looking at our current, past, and projected feedstock characteristics, and redesigning our solids collection system for maximum efficiency over an extended period of time.

Changes to date include the following: timing of automated valves, temperature of collection components, flow path of pneumatically conveyed solids, velocity of pneumatically conveyed solids, ease of maintenance of solids collection system, safety while performing maintenance on solids collection system, procedural changes, and early indicators of malfunctions within the collection system.

Modifications performed have resulted in vastly reduced down time due to solids collection issues. Since these modifications, the system has produced nearly 500 gallons of bio-oil and collected more than 1,000 lbs. of solids without a single solids collection related shut down. Future work will focus on optimal cyclone design and may assess benefits of additional solids collection strategies as we move towards catalytic fast pyrolysis.

500668 Advances of Syngas Chemical Looping Process at the Ohio State University: Pilot Demonstration and Advanced Control Architecture Development

Abstract:

For the past two decades, The Ohio State University (OSU) has been advancing chemical looping technologies as a promising alternatives to conventional processes for chemical and power production with greater process efficiency and carbon emissions reduction. The unique design of counter-current and co-current moving bed can efficiently convert various types of fuel to electricity and/or high value chemicals with nearly 100% CO_2 capture. The presented paper summarizes the most updated development and demonstration results of two syngas chemical looping (SCL) test facilities at OSU â€“ the 250 kW_{th}-3 MW_{th} pilot plant and 25 kW_{th} sub-pilot unit. Specifically, the SCL pilot plant converts syngas produced by coal gasification and generates sequestration-ready COÂ
H_2 with minimal particle circulation. Key design features and experimental data will be discussed. The sub-pilot unit is implemented with an advanced control architecture that enables autonomous start-up, steady state operation and shut-down. The developed hybrid controller, which consists of high level controllers (HLCs) and sliding mode controllers (SMCs), significantly reduces operator intervention requirements during the entire operation. The capabilities of the controller, including disturbance rejection, stage recognition and designed state transition will be presented.

499734 Design of a Continuous Laboratory Setup

Abstract:

Using continuous instead of batch processing has transformed the way we conduct chemistry in the laboratory. As a result the design of continuous setup is a key focus that triggers the analysis of many topics such as project requirements, company strategy, selection of reactors and suitable auxiliaries as well the setup requirements. Characterization of setup via modeling in terms of the flow dynamics as well heat and mass transfer has also an important role to improve and speed up the process development, and allow for a more effective scaling-up.

This work presents the rational followed by the continuous manufacturing group at Hovione to design a mobile, compact and modular setup able to handle homogeneous reactions using coil reactors, demonstrating the lessons learned towards every step during practical application. This work also aims to address on which units operations and reaction types are granted by different modules and show the characterization of coil reactors regarding residence time distribution and heat transfer.

501484 A Novel Process for Continuous Magnesium Metal Production By Carbothermal Reduction of Magnesium Oxide

Abstract:

A novel process for continuous magnesium metal production by carbothermal reduction was designed and fabricated for testing as an efficient alternative to current silicothermic reduction methods. A complete ore-to-ingot process was demonstrated on a small pilot scale (up to 150 g/hr). Magnesium vapor was continuous generated using a fix bed gasification process at 1550Â°C and 1 â€“ 10 kPa in which precursor carbon and magnesia pellets were nearly completely gasified (95 wt%). The rate of magnesium production was inferred from the axial velocity of the bed and the concentration of by-product carbon monoxide in the effluent. Continuous condensation of magnesium vapor was achieved by passing product gases over a moving bed of stainless steel particles (3mm) which absorbed the latent heat of magnesium condensation and the sensible heat of the gas stream. Condensed magnesium and stainless steel particles were removed by gravity and carbon monoxide was removed by pumping. To separate magnesium and steel a vacuum distillation was performed after which product magnesium was collected, melted, and cast into an ingot. Reactor yield based on the product carbon monoxide signal was 70-80% while total ore-to-ingot yield was only 29%. Magnesium oxidation in air prior to vacuum distillation was the likely cause of the discrepancy.

494958 A Pilot-Scale Study of Electrocoagulation for Phosphorus Separation in Liquid Animal Manure

Abstract:

Pork production is one of the most important agricultural activities in the United States, and 75% of the swine operations are geographically located in the Upper-Midwest. There is 20.79 thousand m^3/yr of flushed manure or 2.18 thousand m^3/yr of deep-pit manure produced annually in a typical 2400-head barn. It is a big problem for farmers to handle such a large amount of manure. Swine manure contains a high level of phosphorus (P), ranging from 4.1 g/kg to 18.3 g/kg-dry matter for dairy manure and even higher (9.9 g/kg to 22.7 g/kg-dry matter) for swine manure. The fine particles in manure usually cannot be effectively removed through physical solid-liquid separation methods, such as natural sedimentation, centrifugation, or drum screen, resulting a substantial amount of P contained in these solid fine particles remaining in the liquid manure. Improper land application of manure will create P overload and accumulation in the soil, and eventually leaching to nearby water body, causing eutrophication. Our research group has evaluated electrocoagulation (EC) for P separation from both dairy and swine manure in lab scale, and reached the P removal efficiency of 99% and 85%, respectively. In this work, pilot-scale EC study was conducted with 20 L manure per run and 4-6 runs per day followed by natural sedimentation for 24 h. The results show that 70-90% of P could be concentrated in sludge with a volume of only 5% to 10% of the untreated manure volume. The resulting supernatant is more nitrogen concentrated with a N: P ratio between 10 and 40, compared with the initial ratio of around 2. The sludge has the P contents of between 5% and 10%. Based on these results, we predict that in a typical barn with 2400-head of pigs, operating EC in a deep-pit manure management system will add $0.97 per grown-finish pig. The pilot scale demonstration experiment provides strong evidence that EC has great potential for real application in both small and large barns.

500591 Technical Economic Analysis of an Intensified Integrated Gasification Combined Cycle Plant Design Featuring Membrane and Adsorptive Reactors

Abstract:

Currently, coal-based power plants are the largest source of electricity both globally and in the United States, and despite there being an abundant supply of low-priced natural gas, the Energy Information Administration (EIA) projects that by 2040, coal-based power plants will still provide 254.1 GWe of the generation capacity of the United States, and will account for approximately 36% of the total electricity generation capacity globally1,2. For this reason, coal-based power plants have begun considering the adoption of carbon capture technology in order to reduce the quantity of CO2 emissions, since each year in the United States 6,870 million metric tons of greenhouse gases (GHG) are emitted into the atmosphere, and CO2 represents over 80% of these GHG emissions3.

Because of the environmental concerns these emissions evoke, several carbon capture technologies have been proposed and integrated into these coal power plants with the goal to reduce the CO2 emissions by at least 90%. The main process utilized for carbon separation and capture employs absorption with amine-type solvents, whereby CO2 is separated through the use of a Dual-Stage Selexol unit4.

In this work, the design of an Integrated Gasification Combined Cycle (IGCC) plant featuring membrane and adsorptive reactors is first developed and then analyzed from an economic viewpoint. The design features membrane and adsorptive reactors that have over 90% carbon capture capability without the use of a Dual-Stage Selexol unit. The resultant intensified design is able to produce power more efficiently than traditional IGCC plants with Carbon-Capture Storage (CCS) technology. The UNISIM software (Honeywell trademark) is used to create an intensified flowsheet for the proposed power production plant design, and heat integration studies are presented. A technical economic analysis (TEA) is carried out on the proposed flowsheet to determine economic feasibility of the novel design. The economics of the intensified designs will be compared to a traditional IGCC plant with CCS to identify the extent of process intensification over the baseline design.

1. Annual Energy Outlook 2013, Department of Energy, U.S. Energy Information Administration, 2013.
2. Annual Energy Outlook 2014, Department of Energy, U.S. Energy Information Administration, 2013.
3. Inventory of U.S. greenhouse gas emissions and sinks: 1990-2014. Washington, DC: U.S. Environmental Protection Agency, n.d. Print.
4. Department of Energy/National Energy Technology Laboratory. Cost and Performance Baseline for Fossil Energy Plants Volume 1b: Bituminous Coal (IGCC) to Electricity Revision 2b â€" Year Dollar Update. Document Number: DOE/NETL- 2015/1727.

501648 An Electrochemical Method to Remove Aqueous Sulfide from Swine Manure

Abstract:

Sulfide is synthesized by sulfur reducing bacteria during the anaerobic degradation of swine manure. The year-long pit storage of swine manure allows the accumulation of sulfide, which eventually results in a high risk of hydrogen sulfide (H_2S) releasing to the air during the manure pumping. H_2S is an extremely hazardous and toxic gas; and a number of casualties have been reported due to H_2S poisoning in the past several years. In order to minimize the sulfide level in swine manure and create a safe working environment, an effective way is presented in this study. By applying a certain voltage onto low carbon steel electrodes partially immersed in the swine manure storage, a 90% or above sulfide removal can be achieved after 1 to 2 weeks treatment. In our experiment, the optimal voltage was investigated to be 0.7 V, which allowed the majority removal of sulfide, and meanwhile a guarantee of less electrode material consumption. The effects of potential parameters, including manure solid content, electrode surface area, distance between electrodes, treatment temperature, biogas composition in the reactor headspace, and sulfide recovery after shutting down the power, were investigated. This electrochemical treatment has shown a significant aqueous sulfide removal from swine manure. Optimization based on mentioned parameters can further improve the treatment feasibility in reality.

499078 Pilot-Scale Testing of Electrochemical Removal of Hydrogen Sulfide in Deep-Pit Swine Manure Storage

Abstract:

The presence of hydrogen sulfide (H_2S) is a widespread issue in organic waste storage, and this toxic air contaminant causes serious concerns over environment, health and safety. One example is the deep-pit manure storage, a common practice for swine facilities in upper-Midwestern states. It occasionally experiences human or animal fatal overexposure to manure gas that contains H_2S. In this study, we developed an electrochemical-based oxidation method to remove or greatly reduce sulfide generation in the stored liquid manure. The treatment system needed some equipment installation and retrofit of the current storage pit with little maintenance requirement, low electricity consumption, and low chemicals usage. Lab-scale experiments were conducted in simulated swine manure storage, and systematically analyzed sulfide removal kinetics at varying total solids contents, initial sulfide concentrations, pH, applied voltages, electrode surface areas, electrode distances, and temperatures. An on-farm pilot-scale study was then performed to further confirm the treatment performance in close-to-real environment, and provide necessary data for techno-economic assessment. An electrochemical treatment device was designed and installed in two manure pump-out accesses in a testing barn that held about 200 growing-finishing pigs. Each pit access was divided to two compartments, one for control and the other for treatment installation. Perforated sheets of two electrode materials of low carbon steel (LCS) and stainless steel Type AISI 304 (SS) were installed in accesses and tested in 97 daysâ€™ operation. From the monitored H_2S concentration in pit headspace during the treatment, it can be seen that the LCS treatment decreased a substantial amount of H_2S concentration, achieving removal efficiency of 56%-78% depending on the applied voltage of between 0.6 and 0.7 V. SS treatment removed 13%-38% of H_2S. LCS treatment required a lower voltage application but achieved better removal performance, and is more economically feasible compared to SS installation according to techno-economic assessment. However, electrode material cost accounted for 76%-87% of the net present cost of the treatment assuming 20 yrsâ€™ life-time of the treatment device. Cheaper material, lower surface area, or more effective material will have to be optimized in order to decrease the treatment cost.

495035 Economic and Environmental Impact Analyses of Integrated Dehydrogenation and Hydroformylation in Gas-Expanded Liquid Media

Abstract:

Economic and Environmental Impact Analyses of Integrated Dehydrogenation and Hydroformylation in Gas-Expanded Liquid Media

Dupeng Liu,[1,2] Raghunath V Chaudhari[1,2] and Bala Subramaniam[1,2]

[1]Department of Chemical & Petroleum Engineering, [2]Center for Environmentally Beneficial Catalysis, University of Kansas

Lawrence, KS 66047

Increased shale gas production in the U. S. has also enhanced the availability of C_2-C_5 alkanes (the so-called natural gas liquids) as collateral products, and opportunities for their utilization in novel ways. For example, propylene production capacity via propane dehydrogenation (PDH) is slated to grow rapidly over the next several years.[1] However, for a typical PDH process, the per pass conversion is only about 50%. The separation of propane and propylene mixtures is highly energy and capital intensive. Previously, we had reported the beneficial effects of propane expanded liquids (with propylene/propane mixtures as feed) for propylene hydroformylation with Rhodium-based catalysts. At sufficiently high pressures (>10 bar), the pressurized propane/propylene mixture forms a gas-expanded liquid phase that promotes regio-selective hydroformylation of propylene with homogenous Rh complex. In addition, compared with conventional propylene hydroformylation process to produce butyraldehyde using polymer grade propylene as feedstock, the feed in the propane-expanded liquid (PXL) process is 50% cheaper.[3]

In this talk, economic and environmental analyses of the integrated dehydrogenation and hydroformylation in PXL process are presented and benchmarked against simulated conventional hydroformylation processes for which data were obtained from the Ruhrchemie/Rh™ne-Poulenc and Dow processes.[4-5] Preliminary simulation results indicate that capital investment and production cost associated with the PXL process is lower than the conventional processes (by up to 40% and 25%, respectively) due to the elimination of the C_3 separation section. Life cycle analysis (LCA) was also performed based on plant scale simulation of conventional and PXL processes. These analyses have provided guidance in catalyst design and in choosing materials and operating conditions that favor process economics while lessening environmental footprint.

1. Nawaz Z. Light alkane dehydrogenation to light olefin technologies: a comprehensive review[J]. Reviews in Chemical Engineering, 2015, 31(5): 413-436.

2.Nawaz Z. Dynamic modeling of catofin¨ fixed-bed iso-butane dehydrogenation reactor for operational optimization[J]. International Journal of Chemical Reactor Engineering, 2016, 14(1): 491-515.

3. Liu, D., Chaudhari, R.V., Subramaniam, B. Kinetic Modeling of Propene Hydroformylation in Propane-Expanded Liquid with Rh-Based Complexes, 2016 AIChE annual meeting San Francisco, CA.

4. Wattimena F, Heijmen H J. U.S. Patent 4,521,630[P]. 1985-6-4.

5. Devon T J, Phillips G W, Puckette T A, et al. U.S. Patent 5,332,846[P]. 1994-7-26.

492359 Bio-Energy with Carbon Capture and Storage (BECCS): Are Inefficient Power Plants a Better Option?

Abstract:

Negative emissions technologies (NETs) in general, and bio-energy with carbon capture and storage (BECCS) in particular, have been identified as key technologies in the context of limiting climate change to no more than 1.5 – 2°C (Fuss *et al.*, 2014). By actively removing CO_2 from the atmosphere, NETs provide a service, and thus investors would expect an appropriate risk-adjusted rate of return, varying as a function of the quantity of public money involved. In addition to providing negative emissions, BECCS facilities also generate firm low carbon power. However, as the electricity sector is integrating an increasing share of intermittent renewable energy (iRE), such as wind and solar, the dispatch load factor of such BECCS facilities may be small relative to their capacity. This has the potential to significantly underutilise these assets for their primary purpose of removing CO_2 from the atmosphere. In this study, we present a techno-economic-environmental evaluation of BECCS plants with a range of operating efficiencies, considering their full- and part-load operation relative to a national-scale annual CO_2 removal target, electricity price and negative emissions credit. We find that in all cases, a lower capital cost, lower efficiency BECCS plant is superior to a higher cost, higher efficiency facility from both environmental and economic perspectives. We show that it may be preferable to operate the BECCS facility in base-load fashion, constantly removing CO_2 from the atmosphere and dispatching electricity in a demand-following manner. We show that the use of this "spare capacity" to produce hydrogen for, *e.g.*, injection to a natural gas system for the provision of low carbon heating, can add to the overall environmental and economic benefit of such a system. The sensitivity of the model to biomass carbon footprint is also presented. The only point where our conclusions are altered is where the CO_2 emissions associated with the biomass supply chain are sufficiently large so as to eliminate BECCS CO_2 removal (Mac Dowell & Fajardy, 2017).

References

Fuss, S., Canadell, J. G., Peters, G. P., Tavoni, M., Andrew, R. M., Clals, P., Jackson, R. B., Jones, C. D., Kraxner, F., Nakicenovic, N., Le Quere, C., Raupach, M. R., Sharifi, A., Smith, P. & Yamagata, Y. (2014). Betting on negative emissions. Nature Climate Change, 4 (10), 850–853. https://doi.org/10.1038/nclimate2392

Mac Dowell, N., & Fajardy, M. (2017). Inefficient power generation as an optimal route to negative emissions via BECCS. Environmental Research Letters, at press: https://doi.org/10.1088/1748-9326/aa67a5

APPLICATION OF SOME PREDICTIVE THERMODYNAMIC MODELS FOR INDUSTRIALLY IMPORTANT SYSTEMS

Yizu Zhu
Mohammad Shafiei

AIChE Annual Meeting, Minneapolis, MN

October 31, 2017

CHEMISTRY THAT MATTERS™

THERMODYNAMICS CRITICAL TO INTEGRATED PROCESS ENGINEERING & ECONOMICS

Physical Properties play very important roles in sustainable process engineering and in improving companies' bottom line.

Critical phase equilibrium data are unavailable for compounds under many circumstances:

- Design new products
- Screen conceptual design processes
- Scaleup and commercialize new technologies
- Trouble-shoot & optimize manufacturing plants
- Support techno-economics

Reliable "Predictive Equation Of State" models can be very helpful in tackling the above challenges.

No. 1

IMPACT OF PHYSICAL PROPERTIES ON SYSTEM ECONOMICS

A Close Boiling System

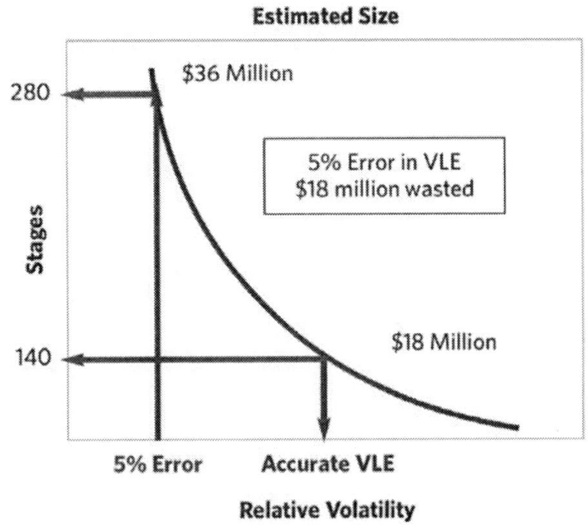

Dhole, et al, Aspen Tech White Paper, 2012

IMPACT OF PHYSICAL PROPERTIES ON SYSTEM ECONOMICS

Ammonia Reaction System using Aspen Plus V8.8

	FEED	PRODUCT		
		"Best available" model with parameters regressed from Data	Using A+ PSRK Prediction	Using Ideal Gas Model
Temperature C	450	450	450	450
Pressure bar	90	150	150	150
Vapor Frac	1	1	1	1
Mole Flow kmol/hr	4	3.272	3.263	3.3
Mass Flow kg/hr	34.06	34.06	34.06	34.06
Volume Flow cum/hr	2.67	1.36	1.36	1.32
Enthalpy Gcal/hr	0.012	0.003	0.003	0.003
Mass Flow kg/hr				
NH3		12.39	12.55	11.92
H2	6.05	3.85	3.82	3.93
N2	28.01	17.82	17.69	18.21
Mass Frac				
NH3		0.364	0.368	0.350
H2	0.178	0.113	0.112	0.115
N2	0.822	0.523	0.519	0.535
Difference%		0.00%	1.28%	-3.82%

The productivity impact is ~17.5$MM/year for a 1,000KTA Ammonia plant (Feb 2017 retail price).

THERMODYNAMIC CHALLENGES AND PREDICTIVE MODELS

- Industry presents challenges for representing
  - Chemically non-ideal systems
  - Involving supercritical components
  - In multiphase regions at extreme temperature and pressures
  - Specialty chemicals and other systems of rare compounds with no VLE-LLE data

- Equation Of State approach is good for high pressure, but until recently, is not-applicable to highly non-ideal chemical systems.

- New Mixing Rules for EOS emerged in the 1990s to represent the interactions in liquid phases of complex mixtures:
  - Predictive SRK (Holderbaum and Gmehling, 1991)
  - Modified Huron Vidal (MHV2) (Dahl, Fredenslund, Rasmussen, 1991)
  - Wong-Sandler (1992)
  - Etc.

No. 4

PENG - ROBINSON EOS

$$P = \frac{RT}{V_m - b} - \frac{a}{V_m(V_m + b) + b(V_m - b)}$$

Mixing Rule: $a = \sum_i \sum_j x_i x_j (a_i a_j)^{0.5}(1 - k_{ij}), \quad b = \sum_i x_i b_i$

$$a_i = \alpha_i 0.45724 \frac{R^2 T_{ci}^2}{P_{ci}}, \quad b_i = 0.07780 \frac{RT_{ci}}{P_{ci}}$$

$$\alpha_i(T) = [1 + m_i(1 - T_{ri}^{0.5})]^2, \quad m_i = 0.37464 + 1.54226\,\omega_i - 0.26992\,\omega_i^2$$

Binary interaction parameter k_{ij} needs to be regressed from phase equilibrium data

No. 5

APPLICATION OF PREDICTIVE EOS MODELS FOR HIGH PRESSURE AND HIGHLY NON-IDEAL SOLUTION SYSTEMS

SRK: $P = RT/(v-b) - a/v(v+b)$

PSRK's Mixing Rule:

$$b = \sum x_i b_i$$

$$a = b[g^E/A_1 + \sum x_i a_i/b_i + (RT/A_1)\sum x_i \ln(b/b_i)]$$

g^E : Excess Gibbs Energy from UNIFAC Group Contribution Activity Coefficient Model

A_1 : constant

(Holderbaum and Gmehling, 1991; Dahl and Michelson, 1990)

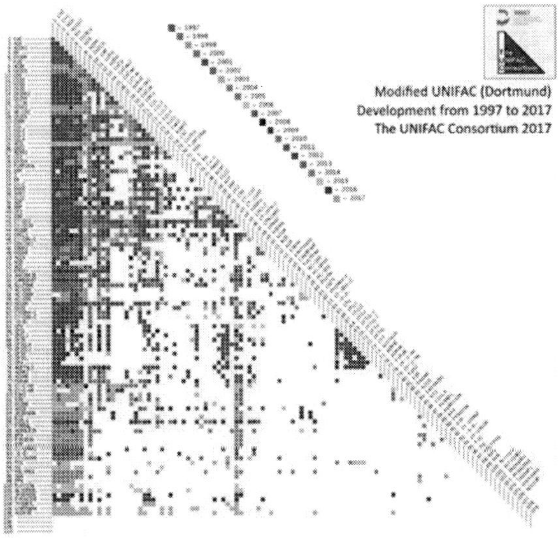

No. 6

KEY FEATURES OF THE NEW MIXING RULES

- A composition-dependent term F(x)
- F(x) is related to Excess Gibbs Energy or Helmholtz Energy
- Allows a single Equation Of State model to represent the phase behavior of:
  - Low Density Fluid (usually by EOS)
  - High Density Fluid (usually by Activity Coefficient Models)
  - Polar Fluid including highly non-ideal LLE systems

No. 7

PREDICTION FOR MEOH-H2S PHASE BEHAVIOR

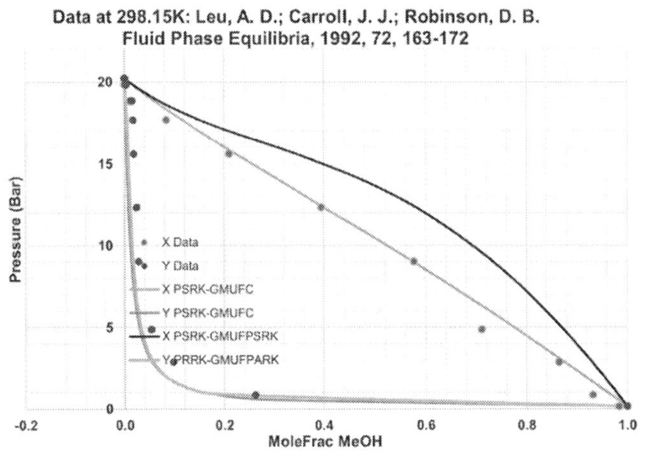

No. 8

METHANE-N-BUTANE P-XY AT HIGH PRESSURE

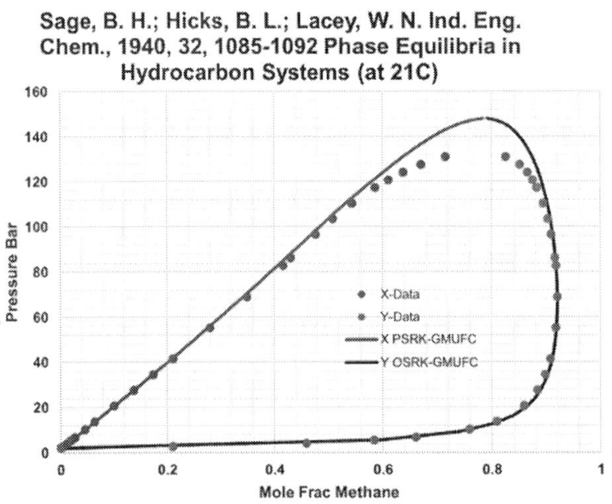

No. 9

EtAc–HEXANE SYSTEM T-XY PHASE DIAGRAM

EtAc-Hexane System at 1 atm
Fernandez, L.; Perez, E.; Ortega, J.; Canosa, J.; Wisniak, J.
J. Chem. Eng. Data, 2010, 55, 5519-5533

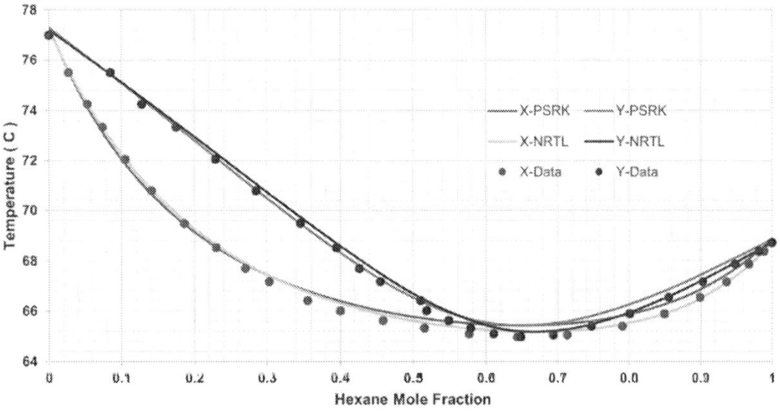

No. 10

PSRK PREDICTION FOR HIGH P, STRONGLY NON-IDEAL SYSTEMS

MEOH-CO2 SYSTEM

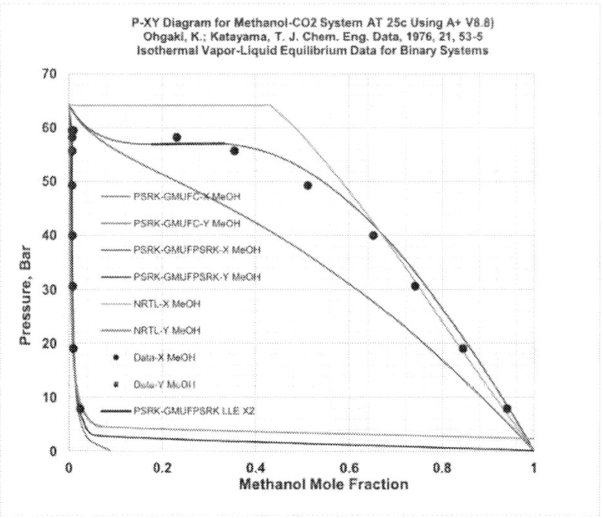

No. 11

CO2-CYCLOHEXANE SYSTEM AT HIGH PRESSURE

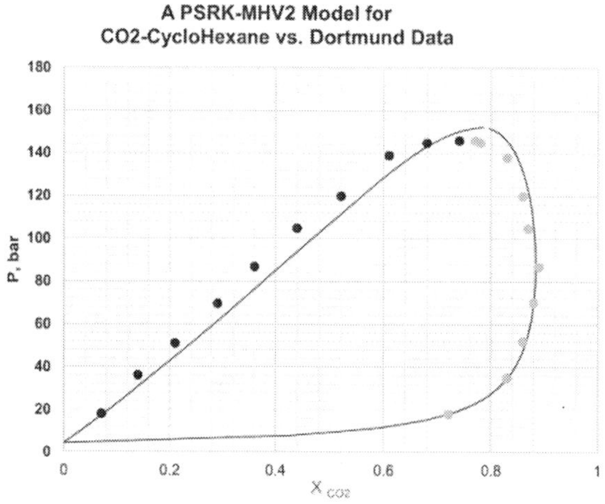

No. 12

NH3-WATER SYSTEM AT TYPICAL PLANT OPERATION CONDITION

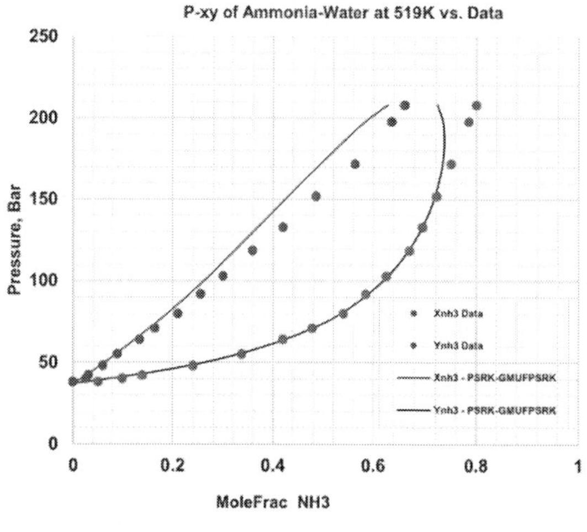

No. 13

N-BUOH – WATER HETEROGENEOUS AZEOTROPE PREDICTION

Liadosa, E.; Monton, J. B. ; Burguet, M. C.; Munoz, R. J. Chem. Eng. Data, 2008, 53, 108-115 Phase Equilibrium Data (101300 N/m2)

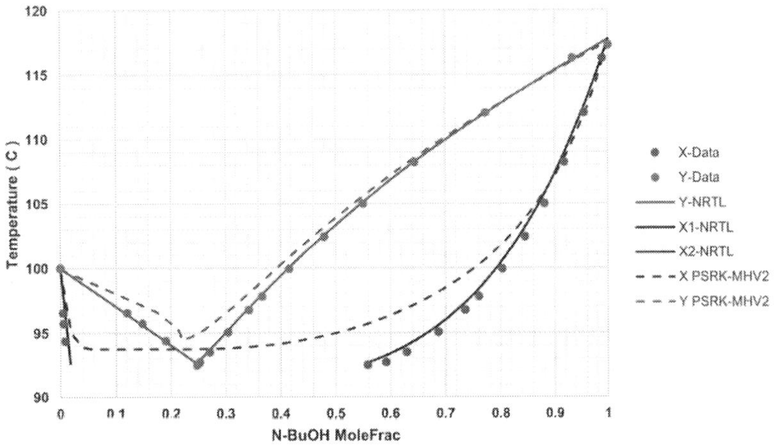

No. 14

SEPARATION IN AN ETHYLENE PLANT WITH PSRK

	Tower Reflux	Tower Bottom Liquid Rate	Tower Top Temperature
	t/h	t/h	C
Plant Data	12.00	52.30	-98.50
RKS	12.31	51.00	-100.10
PSRK	12.99	52.40	-99.50

No. 15

TEMPERATURE IMPACT ON HYDROCARBON CONCENTRATION IN RESIN USING POLYSRK FOR PRODUCT QUALITY IMPROVEMENT AND CAPACITY INCREASE

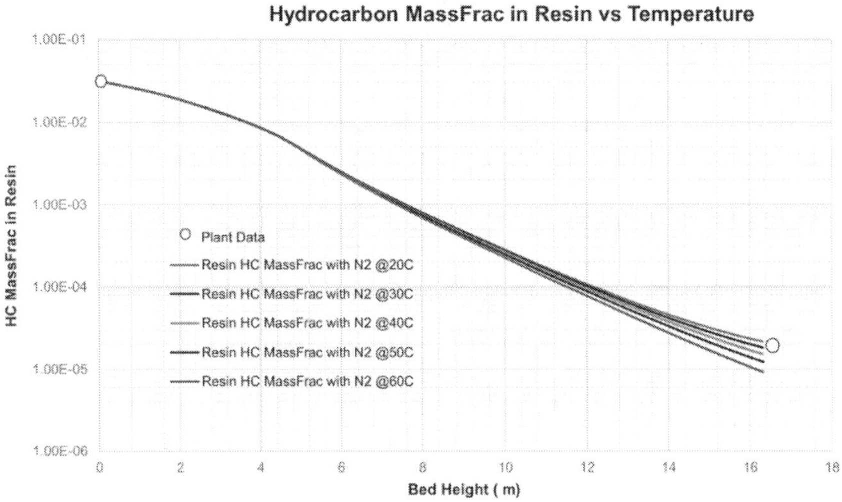

No. 16

CONCLUSIONS

- Physical properties have critical impact to the economics of process design and manufacturing operations.
- Predictive Equation of State models provide a powerful tool in representing highly non-ideal systems under high pressure where data are unavailable and other types of thermodynamic models are unable to handle.
- Predictive EOS models can be used with Excess Gibbs Free Energy of various liquid activity coefficient models available in simulation software such as Aspen Plus where binary interaction parameters are available (such as NRTL, WILSON, UNIQUAC, etc).
- This provides a powerful tool to handle a variety of complex systems where data are sparsely available:
  - Conceptual design of new products and processes
  - R&D
  - Scale-up and commercialization of innovative technologies
  - Manufacturing retrofitting and performance optimization

No. 17

DISCLAIMER

DISCLAIMER: THE MATERIALS, PRODUCTS AND SERVICES OF SAUDI BASIC INDUSTRIES CORPORATION (SABIC) OR ITS SUBSIDIARIES OR AFFILIATES ("SELLER") ARE SOLD SUBJECT TO SELLER'S STANDARD CONDITIONS OF SALE, WHICH ARE AVAILABLE UPON REQUEST. INFORMATION AND RECOMMENDATIONS CONTAINED IN THIS DOCUMENT ARE GIVEN IN GOOD FAITH. HOWEVER, SELLER MAKES NO EXPRESS OR IMPLIED REPRESENTATION, WARRANTY OR GUARANTEE (i) THAT ANY RESULTS DESCRIBED IN THIS DOCUMENT WILL BE OBTAINED UNDER END-USE CONDITIONS, OR (ii) AS TO THE EFFECTIVENESS OR SAFETY OF ANY DESIGN OR APPLICATION INCORPORATING SELLER'S MATERIALS, PRODUCTS, SERVICES OR RECOMMENDATIONS. UNLESS OTHERWISE PROVIDED IN SELLER'S STANDARD CONDITIONS OF SALE, SELLER SHALL NOT BE RESPONSIBLE FOR ANY LOSS RESULTING FROM ANY USE OF ITS MATERIALS, PRODUCTS, SERVICES OR RECOMMENDATIONS DESCRIBED IN THIS DOCUMENT. Each user is responsible for making its own determination as to the suitability of Seller's materials, products, services or recommendations for the user's particular use through appropriate end-use and other testing and analysis. Nothing in any document or oral statement shall be deemed to alter or waive any provision of Seller's Standard Conditions of Sale or this Disclaimer, unless it is specifically agreed to in a writing signed by Seller. Statements by Seller concerning a possible use of any material, product, service or design do not, are not intended to, and should not be construed to grant any license under any patent or other intellectual property right of Seller or as a recommendation for the use of any material, product, service or design in a manner that infringes any patent or other intellectual property right.

THANK YOU AND Q&A

Towards Bio refinery Process Integration in Oil Refineries

Harvey Arellano-Garcia*, Elham Ketabchi, Tomas Ramirez Reina

Department of Chemical and Process Engineering, University of Surrey, Guildford, GU27XH, United Kingdom

*h.arellano-garcia@surrey.ac.uk

Abstract

An investigation into the integration of bio-refineries with oil refineries is considered in this study. The identified possible routes of integration will be discussed in this contribution. For this purpose, the different proposed alternatives and their configurations were simulated and analysed. These flow schemes formed the basis upon which a series of simulations in CHEMCAD™ and Aspen plus® were undertaken. The developed models simulated key integrations e.g. a gasification unit that is fed from pyrolysis oil, biodiesel, and refinery residue, before being combined into one system involving all three. Varying forms of synthesis for these three feeds were also considered, focusing on novel techniques as well as environmentally friendly options that made use of waste products from other processes. The simulations found that the resulting flow from the gasification unit contained the expected composition of H_2 and CO_2, with CO in slight excess. The H_2:CO ratio was also found to be adequate enough for further processing in the form of producing hydrogen (Water Gas Shift reaction) and liquid hydrocarbons (Fischer-Tropsch). This study also found that through the manipulation of oxygen content, the H_2:CO ratio was improved. Changes in composition of the output from the gasification unit were also evaluated through influencing the input flows from the three sources mentioned earlier. Furthermore, through the use of low value streams and recycled waste, as well as the use of non-fossil based fuels, this integration drastically reduces the negative environmental impact that the refinery has as well as preventing excessive wastes of energy, while not increasing costs for advanced equipment or pre-processing.

Keywords: Process integration, bio-refinery, oil refinery, GHG, waste reduction.

1. Introduction

The main idea is to create a flow sheet of integrated units and processes of an oil refinery with a bio-refinery. The flow sheet construction has been inspired by the work of Schinski (2014). This integration is appealing as bio-refinery feed stocks are more readily available due to the wide variety of biomass that can be used, i.e. sugar and wood that produce bio-fuels, pyrolysis oil in this study, of comparable quality to fossil fuels and therefore, energy. Furthermore, bio-diesel produced from waste water treatment as an extra added-value product has been considered (Kiracofe, 2010).

These processes mentioned including oil refinery residue, oxygen and steam are all inputs for the gasification unit in order to produce syngas.

Once the syngas has been purified, it enters the WGS and FT reactors to produce fuel (liquid hydrocarbon) and hydrogen. Energy obtained from these processes can also be used in a combined heat and power (CHP) unit to maximise the process integration.

The tail gas outlet from the FT reactor enters a separation unit, separating the carbon monoxide from the stream. The remainder of which, enters the refinery again.

Pyrolysis oil used in this study is specifically fractionated butanol that comes from sugar fermentation. This is more favourable as a bio-fuel due to the need of mitigating greenhouse gas (GHG) emissions with the use of developing technologies, when compared to conventional fuel. Figure 1 shows the flow sheet used in this work for modelling and simulations in CHEMCAD™ and Aspen Plus®. Moreover, sensitivity analysis was carried out to determine sensible output changes by varying several input properties.

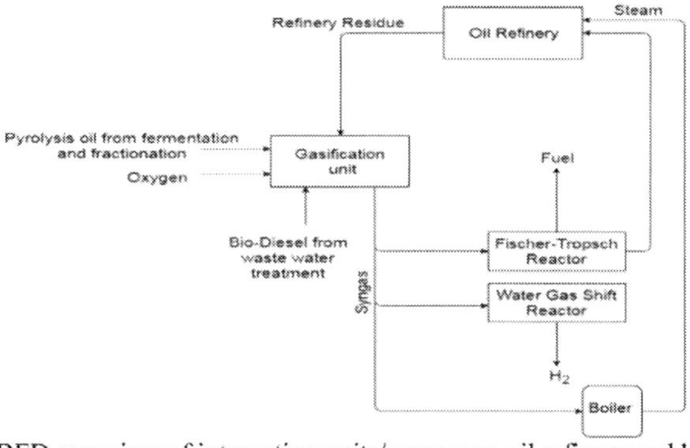

Figure 1. BFD overview of integrating units/processes: oil refinery and bio-refinery

The other route considered in this flow sheet is the production of bio-diesel, as seen in the bottom of Figure 1, from the waste water sludge that is produced from the waste water treatment process via trans-esterification, then using this bio-diesel as another input for the gasification unit. The waste water sludge is rich in fatty acids, which are the main components required for trans-esterification that produces bio-diesel and glycerol. These two products are easily separated through methods such as gravity settling (Kiracofe, 2010).

The gasification unit operating under isothermal and steady state conditions also uses refinery residue, in addition to the bio-diesel and pyrolysis oil, with oxygen and steam to produce syngas, which is the main product, while also producing slag. The use of these three main streams that originate from previously unused material, to create syngas, is to promote more "green" energy production and to offer solutions to dwindling fossil fuel supply.

The conversion of biomass via gasification and hydro-pyrolysis followed by the conversion of natural gas via FT synthesis to liquid fuel is another alternative to these integration pathways proposed by Gencer et al (2014) and is included it in the integrated approach. The tail gas produced from the FT reactor could also be entered in a separator, where carbon dioxide from the remainder gas is separated before being returned to the oil refinery. Another aspect of integration proposed is the conversion of syngas to steam via boiling to produce steam for the CHP plant. The output of the boiler being steam could either be directly used in the refinery operations or could be converted to electricity for the benefit of a refinery.

2. Design and simulation

2.1. Mass and energy balance of the gasification unit

The first step is to identify the specifications of the inputs to the gasification unit being pyrolysis oil, biodiesel, and refinery residue. It is considered that the product of a conventional pyrolysis unit with the efficiency of 20 % is 0.956 kg/s using information provided in the work by Desai (2011). The common composition of pyrolysis oil is obtained from Vivarelli and Tondi (2004). The bio-diesel used as an input in this flow sheet is obtained from the process of using waste from waste water treatment as mentioned before. The values used for this simulation are derived from Kiracofe's 2010 work as well as simple mass and energy balances leading to the values demonstrated in the caption for Fig 2. The composition of bio-diesel chosen for this work is derived from the work of Hoekman et al.(2012). The amount of bio-diesel shown here is minimal due to conversion processes only having yields of maximum 13 % as shown in Kiracofe's (2010) research. Considering the input volumes to an oil refinery, we can assume that the outlet volume in the form of refinery residue will be of the order of $1x10^4$ kg/h. In this case, the refinery residue was initially chosen to be 60,000 kg/h for use in the simulation. Standard stoichiometric approaches are not applicable for biomass processes due to the fact that they are not composed of well-defined chemical species, therefore, the amount of oxygen required for the gasification process was derived from the calculation of the empirical formula in (Bernocco et al., 2011). Moreover, through the use of simple energy balance, the temperature of the oxygen and steam mixture required for the gasification unit was calculated.

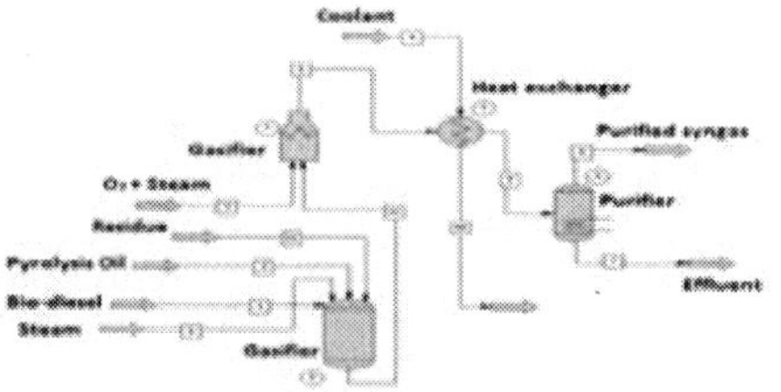

Figure 2. CHEMCAD™ Version 6.5.5 setup for gasification and purification (gasification conditions: 40 bar, 1400 °C, steam temperature 350 °C and 99.6 % O_2 concentration – adapted from Chao 2015)

The simulation was improved, however, by reducing the amount of oxygen for the gasification process in order to obtain higher quality syngas, with a more desirable ratio of $CO:CO_2$, in favour of CO.

2.2. Mass and energy balance of the water gas shift reactor

Following the gasification modelling (Fig. 1), the next step is the water gas shift reactor. The produced gas or the purified syngas enters the WGS reactor in order to produce hydrogen rich syngas. For this reaction, an equilibrium reactor is selected for the WGS reaction due to the chemistry involved dictating the reliance on equilibrium to favourably produce hydrogen.

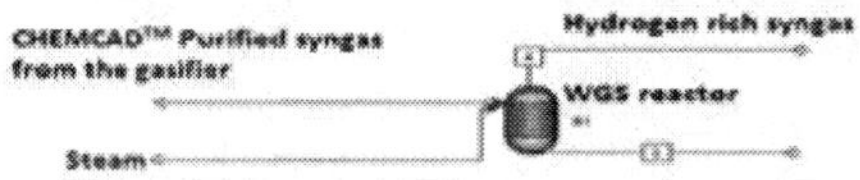

Figure 3. Simulation setup for WGS reaction in Aspen Plus® Version 8.6 (WGS reactor conditions: 36.8 bar, outlet temperature 612 °C, Steam temperature 350 °C, Steam pressure 165 bar – adapted from Kopyscinski et al, 2010)

3. Results and Discussion

The composition products of the WGS reactor are presented in tables 1 and 2. The values demonstrate high amounts of carbon monoxide followed by hydrogen and carbon dioxide with the percentages of approximately 4 %, 95 %, and 1 %, respectively. The obtained composition is as expected, showing the values for good quality syngas which has been improved even more by conducting sensitivity analyses. It must be noted that the amount of methane produced in the gasification unit were negligible. As can be seen in table 2, the WGS reaction products obtained, result in having the desired ratio of 2.0105 being suitable for further processing such as FT reactions.

Table 1.Syngas composition obtained from simulation after improvements made through sensitivity analyses

Syngas composition	Amount (kg/h)	Amount (kmol/h)
H_2	3564.34	1768.20
CO	61817.50	2206.98
CO_2	60.33	1.37

Table 2.WGS reaction (WGSR) products derived from simulation

WGSR Product Composition	Amount (kmol/h)
CO	1231.06
H_2	2475.05
H_2O	2376.84
CO_2	827.84

4. Conclusions

The results of the simulation imply that the syngas produced from the gasification process demonstrate the expected composition of CO, H_2, and CO_2 having high amounts of CO.

The results also depict the H_2:CO ratio of the WGSR products to be as expected, and thus, meeting the requirement for further processing. This ratio has been improved through enhancements made in the amount of oxygen, supported by using information obtained through sensitivity analysis. Moreover, changes in composition of the output from the gasification unit were also evaluated by modulating the input flows.

In conclusion, through the use of non-fossil based fuels as well as low-value streams and by recycling waste, this integration method will be able to decrease the negative environmental impacts of conventional oil refining while preventing excessive wastes of energy without increasing costs for advanced equipment or pre-processing.

Finally, it could be shown that these integrations will increase the profitability of any refinery that incorporates them properly and effectively while reducing overall refinery wastage.

References

Bernocco, D., Greppi, P., & Arato, E. (2011). *A novel approach to the biomass pyrolysis step and product lumping. Computer Aided Chemical Engineering* (Vol. 29). Elsevier B.V. http://doi.org/10.1016/B978-0-444-53711-9.50023-7

Chao, L. I., Zheng-hua, D. A. I., Ji, Y., Guang-suo, Y. U., & Fu-chen, W. (2015). Modelling and energy analysis of an integrated coal gasification and pyrolysis system for synthetic natural gas. *Journal of Fuel Chemistry and Technology*, *43*(7), 779–789.

Desai, P. D. (2011). Design of pyrolysis reactor. Retrieved June 28, 2016, from https://www.scribd.com/doc/100920538/Design-of-Pyrolysis-Reactor

Gencer, E., Mallapragada, D., Tawarmalani, M., & Lafayette, W. (2014). Synergistic Biomass and Natural Gas Conversion to Liquid Fuel with Reduced CO 2 Emissions. In *e 8th International Conference on FOCAPD* (pp.525–530).

Hoekman, S. K., Broch, A., Robbins, C., Ceniceros, E., & Natarajan, M. (2012). Review of biodiesel composition , properties , and specifications. *Renewable and Sustainable Energy Reviews*, *16*(1), 143–169.

Kiracofe, N. L. (2010). *Converting wastewater treatment facilities into biorefineries: Biodiesel from wastewater microorganisms.* University of Delaware.

Kopyscinski, J., Schildhauer, T. J., & Biollaz, S. M. A. (2010). Production of synthetic natural gas (SNG) from coal and dry biomass – A technology review from 1950 to 2009. *Fuel*, *89*(8), 1763–1783.

Schinski, W. L. (2014). Hybrid refinery for co-processing biomass with conventional refinery streams. United states: United states Patent and trademark office.

Vivarelli, S., & Tondi, G. (2004). *Bioenergy for a sustainable development.* Chile.

501634 Process Intensification in the Synthesis of Tributyl Citrate: Pilot Scale Validation

Abstract:

Over the few last decades, bio based plasticizers have been their market share due to the generalized toxicity concern from phthalate esters and other traditional plasticizers. Tributyl citrate (TBC) may be synthesized from citric acid and n-butanol in presence of an acid catalyst. Current production process is highly energy intensive due to its semi-batch configuration and the excess of butanol used at the reaction stage which must be recovered downstream.

The present work evaluated the production of TBC via reactive distillation and other intensifies configurations to increase energy and mass efficiency and furthermore profitability. The thermodynamic model to describe the non-ideal behavior of the reactive system has been regressed from liquid-liquid, solid-liquid , and vapor-liquid experimental equilibria of binary and ternary mixtures. The kinetic model has been regressed from experiments with homogeneous and heterogeneous catalysts. The conceptual design has been carried out using reactive residue curve maps with the introduction of the substitution degree to visualize multicomponent mixtures on a 2D color-scale. Pilot experimental evaluation was carried out in a 7m tall 3â€ wide stainless steel reactive distillation column available in our laboratory with structured packing similar to KATAPAK SP-11 with Amberlyst 70 as catalyst in the reactive section and 1â€ nutter rings in the stripping and recovery sections. The experimental pressure was 74kPa, the atmospheric pressure in Bogota, Colombia. Variables of study during the pilot experimentation include: reflux ratio, reactive mixture feed, use of pure butanol in a side stream, use of azeotropic grade over dehydrated butanol, and the use of prereactor. Pilot plant data were reconciled to develop a robust model for process up-scaling. An industrial process was modelled and optimized for a production capacity of 1000Ton/yr with a natured inspired algorithm. Results from optimization indicate a citric acid conversion>99.9%, selectivity towards TBC >90%, energy savings of 27% and cost reduction of 13% over the conventional process.

491296 Process Intensification in Multicomponent Distillation

Abstract:

Process intensification (PI) is an emerging concept in chemical engineering which describes the design innovations that lead to significant shrink in size and dramatic boost in efficiency in a process plant. Distillation, which accounts for 90-95% of all separations in a chemical plant and ~3% of the world's energy consumption, is therefore a crucial component in PI. In the context of multicomponent distillation, we believe that PI stands for minimizing the number of pieces of equipment as well as total cost while maintaining high efficiency through innovative process synthesis. This will lead to compact, easy-to-operate, energy efficient and cost effective processes. Here, we consider several different levels of PI in multicomponent distillation, including: 1) Introducing thermal couplings that eliminates some or all reboilers and condensers to save energy and reduce capital cost; 2) Using multi-effect distillation along with converting thermal couplings to liquid-only transfers to further reduce heat duty, enhance energy efficiency, and improve operability of thermally coupled distillation configurations; 3) Performing simultaneous heat and mass integration among thermally coupled distillation columns to reduce the number of columns and heat duty requirement; 4) Conducting any thermally coupled distillation in only one single column shell using the form of a dividing wall column that is fully operable; 5) Being able to systematically derive numerous configurations that use 1 to (−2) column shells from a traditional (−1)-column configuration. Here, we demonstrate these aspects of PI through examples, illustrating the importance of PI in multicomponent distillation and how this concept can be utilized by process engineers to design cost effective and energy efficient distillation systems.

489897 Analysis of Key Metrics in the Use of Oxidative Dehydrogenation for the Production of Ethylene from Shale Gas Ethane

Abstract:

The world-wide market for primary light olefins will approach half a trillion pounds per year by the year 2020. Light olefins are conventionally produced in the US by the steam cracking of hydrocarbons from ethane to gas oil. With the abundance of shale gas components, the feedstock of choice in the US is ethane. The reaction technology is decades old and results in a wide range of byproducts, some valuable, some with fuel value, and some requiring waste treatment before disposal. A techno-economic analysis was performed using reaction data provided by the DOE-INL for Oxidative Dehydrogenation (ODH) and modeling the entire expected processing using ASPEN-HYSYS V. 9.0. A set of performance metrics and a cost of production procedure commonly called a required netback analysis were used to compare this intensified ODH process with conventional steam pyrolysis technology. The new ODH route uses a mixed metal oxide catalyst in packed tubes with a heat transfer fluid in the reactor shell for heat management. Results indicate that the new route has much lower variable cost, fixed costs and capital recovery reaching shutdown economic level. Regarding the 20+ metrics developed, the ODH route shows significant advantages and integrates well with certain novel separations technology currently in the pipeline.

497448 Continuous Manufacturing As an Enabler for Intensified Process Design and an Integrated Control Strategy in Pharmaceutical Synthesis

Abstract:

Development, scale-up, and implementation of a continuous multi-step route for manufacturing of an active pharmaceutical ingredient are detailed. Long recognized as essential for economical process intensification in other chemical industries, continuous manufacturing stands to offer many of the same advantages to the pharmaceutical industry. These include more inherent scalability on account of improved heat and mass transfer, access to more aggressive reaction conditions, smaller footprint equipment sets placing less material at-risk, more opportunities to leverage online measurement and automation, and established control of the process at steady-state. The integration of all of these advantages enables implementation of a control strategy that ensures quality under more cost-effective, and often safer, flow processing conditions.

For the case study herein, the transition from laboratory development to a continuous process demonstration operating at 50â€"100% of the commercial-scale in walk-in fume hoods is described. Scale-up efforts integrating multiple unit operations in flow and leveraging process analytical technology (PAT) are demonstrated to test process robustness and generate process knowledge that directly informs the target conditions and operating ranges for the commercial-scale manufacturing process. In this example, transfer of the process for a late-stage clinical supply campaign within the external manufacturing network is detailed, culminating in a continuous route with two plug flow reactor (PFR) transformationsâ€"an amidation and a high-temperature deprotectionâ€"operating in conjunction with a continuous mixed-suspension, mixed-product-removal (MSMPR) crystallization and parallel filtration with *in situ* solids dissolution. Genotoxic impurity control is afforded through process design and engineering controls and is verified with online and offline analyses. The process is shown to operate in continuous mode for 18 days, affording more than 150 kg drug substance in exceptional purity. These steps will be at the core of a multi-step continuous unit operations sequence in the commercial process.

493712 A Systematic Method for Chemical and Biochemical Sustainable Process Synthesis, Design and Intensification

Abstract:

In this highly competitive world of chemicals, challenges like ever-increasing energy and utility cost, pressure to increase resource efficiency and new environmental regulations [1] requires systematic process synthesis-design solutions. The objective of systematic process synthesis-design is to identify the processing route, which gives a desired product from a specific set of raw materials including unit operations design, utility needs, waste generation and sustainability parameters. However, to find innovative and more sustainable process design solutions, extension of the current methods is necessary that can be generated by process intensification. Process Intensification (PI) provides a pathway to design new and innovative solutions or retrofit of existing ones leading to economic, sustainable and efficient solutions. Thus, it is logical to integrate process synthesis-design and intensification to generate more innovative, sustainable and efficient chemical and biochemical solutions.

In this work, a generic three-stage hierarchical decomposition method based on unit operation, task and phenomenon level is applied for sustainable production of succinic acid. This approach examines several succinic acid processing routes from different feeds including bio (glucose, glycerol etc.) and petro (maleic anhydride) based raw materials that are collected through available data and current technologies reported in literature. Using this information, a generic superstructure of processing routes is created that describes a network of configurations representing multiple designs for the production of succinic acid. The next step is to solve the superstructure in order to find the optimal processing route or flowsheet. This has been done through Super-O [2], a software interface that guides the formulation and optimization of the superstructure to find optimal processing route. Thus, in stage 1, process synthesis is performed to generate an optimal flowsheet among numerous alternatives for succinic acid. In stage 2, the base-case design is developed based on the route selected in stage 1, which is then further analyzed to identify the process hotspots. This is done by first performing simulation to obtain the steady state mass and energy balance data for the selected route. These data are then used for economic analysis, sustainability evaluation [3] and life cycle assessment [4] where a set of calculated indicators are used to identify the bottlenecks or process hotspots. These hotspots are then translated to design targets that are used in the next (3rd) stage where, a phenomenon based synthesis-intensification method is applied [5]. Here, the phenomena are identified by using the PI knowledge base that are representing the tasks performed by unit operations within the base case design. These phenomena are further analysed and re-combined using combination rules to generate new and/or existing unit operations including hybrid operations which in turn are combined to form new and innovative process alternatives. Finally, from the set of feasible intensified process alternatives for production of succinic acid, the best in terms of economic and environmental sustainability is identified.

In this presentation, the implementation of 3-stage method into an integrated multi-scale computer-aided framework will be presented and highlighted by its application to the production of succinic acid to generate more sustainable solutions. The presentation will also discuss the computer aided tools used in this work to achieve the sustainable production of succinic acid.

References:

[1] Chemicals, Retrieved April 4, 2017, from https://ec.europa.eu/growth/sectors/chemicals_en

[2] M. Bertran, R. Frauzem, A. Arcilla, L. Zhang, J. M. Woodley, R. Gani, Computers & Chemical Engineering, 2017, doi: 10.1016/j.compchemeng.2017.01.030.

[3] A. Carvalho, H. A. Matos, R. Gani, Computers & Chemical Engineering, 50 (2013): 8-27.

[4] S. Kalakul, P. Malakul, K. Siemanond, R. Gani, Journal of Cleaner Production, 71 (2014): 98-109.

[5] D. K. Babi, J. Holtbruegge, P. Lutze, A. Gorak, J. M. Woodley, R. Gani, Computers & Chemical Engineering, 81 (2015): 218-244.

489611 Strategies for Process and Size Selection of Natural Gas Liquefaction Processes: Specific Profit Portfolio Approach By Cost Based Optimization

Abstract:

The natural gas liquefaction process is highly energy intensive process due to the cryogenic operating characteristics.[1] Thus, the natural gas liquefaction process has the energy intensive characteristics because of its cryogenic operation condition and most of energy is consumed by compression unit.[2] Therefore, minimizing compression energy requirement is the major concern in design and optimization of the natural gas liquefaction process. The natural gas liquefaction processes can be classified by the number of refrigeration cycles and the type of refrigerants used.[3] Because different processes have different characteristics in energy efficiency and number of equipment, the comparison between various liquefaction processes have to be performed to find the optimal design; in terms of the energy as well as cost Furthermore, the plant size also affects to the capital cost. In economics of scale, larger size of plant can make more profit. The amount of the increased profit by increasing the size is different to different processes. Therefore, the most profitable process can be different by different size of plant. For these reasons, it is needed to optimize in terms of the cost with different liquefaction processes to find optimal design and operating condition in various plant sizes.

The main purpose of this study is to make strategies of process and size selection by finding the optimal natural gas liquefaction processes in various plant sizes. To find optimal design and size, the operating cost including energy requirement and capital cost have to be considered simultaneously. The cost based model can consider the plant size which cannot be considered in thermodynamic model. Moreover, the price and flow rate of the natural gas feed and LNG also affect to the profit. Therefore, the profit optimization model have to be developed because it can consider the cost of law material and the product.

The profit optimization model is developed with energy self-sufficient liquefaction system. Among various natural gas liquefaction processes, Single Mixed Refrigerant (SMR), Dual Mixed Refrigerant (DMR), and Propane-precooled Mixed Refrigerant (C3MR) processes are targeted because they are well known and have representative refrigeration characteristics. Optimization model is developed by gPROMS software and the model includes thermodynamic model and cost model. The cost model basically follows the six-tenth factor rule. The objective function is max. specific profit which is profit per unit mass of LNG (USD/ton-LNG). The unit prices of natural gas and LNG are applied based on the Henry-hub prices announced by EIA, the unit prices of the natural gas and the LNG are applied 2.5 USD per Mcf and 4.5 USD per Mcf respectively to reflect recent prices.[4] The detailed mathematical model basically follows the model which is presented in previous work.[5]

The profit optimization with various size of SMR, DMR, and C3MR processes are performed and the results of optimizations are compared and discussed. The portfolios for the liquefaction ratio, specific energy requirement, specific cost, and specific profit obtained by the profit optimizations for each process. The most economical size range for each process design is figured out by drawing specific profit portfolios. SMR process which has the simplest configuration has an economic strength in the small plant size. Even though the energy efficiency is the lowest among three liquefaction processes, the low plant cost of SMR process can cover the disadvantage of the low efficiency in the small plant size. However, as the plant size increases, DMR process catches up SMR process in terms of the specific profit. The configuration of DMR process is more complex and it requires more equipment than SMR process. For the medium size of plant, the higher energy efficiency makes the DMR process economical compare to SMR process. For the large plant size, the C3MR process is most economical because of the highest efficiency although it requires much more equipment compare to SMR and DMR processes.

This study is performed with 2.5 USD per Mcf for the natural gas price and 4.5 USD per Mcf for the LNG price. By applying those prices, the economical ranges of each process are figured out. In addition, the natural gas liquefaction process selection strategies are suggested. There is a possibility that different prices of the natural gas and the LNG could show different economical ranges of various liquefaction processes. However, the optimizations with the profit optimization model which is developed in this thesis can be also applied to the process selection strategies for different situation. In addition, the specific cost and specific profit portfolios for various process designs by size can be applied to the supply chain network research field, and it can make the result more rigorous by multi-scale problem formation. Moreover, the specific profit portfolio approach can be applied to other chemical process industries.

REFERENCES

[1] **Mortazavi, A.,** *et al.*, "Enhancement of APCI cycle efficiency with absorption chillers," *Energy*, **35** (9), pp. 3877-3882 (2010).

[2] **Lee, I., et al.,** "Decision Making on Liquefaction Ratio for Minimizing Specific Energy in a LNG Pilot Plant," *Ind. Eng. Chem. Res.*, **54** (51), 12920-12927 (2015).

[3] **Lim, W.,** *et al.*, "Current status and perspectives of liquefied natural gas (LNG) plant design," *Ind. Eng. Chem. Res.*, **52** (9), 3065-3088 (2013).

[4] **U.S. Energy Information Administration,** "U.S. Natural Gas Prices," https://www.eia.gov/dnav/ng/ng_pri_sum_dcu_nus_m.htm, (accessed on Apr. 2017).

[5] **Lee, I., and Moon, I.,** "Economic Optimization of Dual Mixed Refrigerant Liquefied Natural Gas Plant Considering Natural Gas Extraction Rate," *Ind. Eng. Chem. Res.*, **56** (10), 2804-2814 (2017)

MULTI-OBJECTIVE OPTIMIZATION OF CRUDE DISTILLATION UNIT, VACUUM DISTILLATION UNIT AND HYDROCRACKING REACTOR IN DIESEL PRODUCTION PROCESS TO MAXIMIZE PROFITABILITY AND REDUCE GREENHOUSE GASES USING NSPSO AND NSGAII BIOINSPIRED TECHNIQUES

PhD Stu. Johana Orjuela
PhD Stu. Camilo Monroy

20/11/2017

AICHE ANNUAL MEETING - 2017

2

CONTENT

- PROBLEM STATEMENT
- OBJECTIVES
- METHODOLOGY
- RESULTS
- CONCLUSIONS

Chemical and Environmental Engineering Department
Universidad Nacional de Colombia

Environmental Engineering
Universidad de Cundinamarca

3

PROBLEM STATEMENT

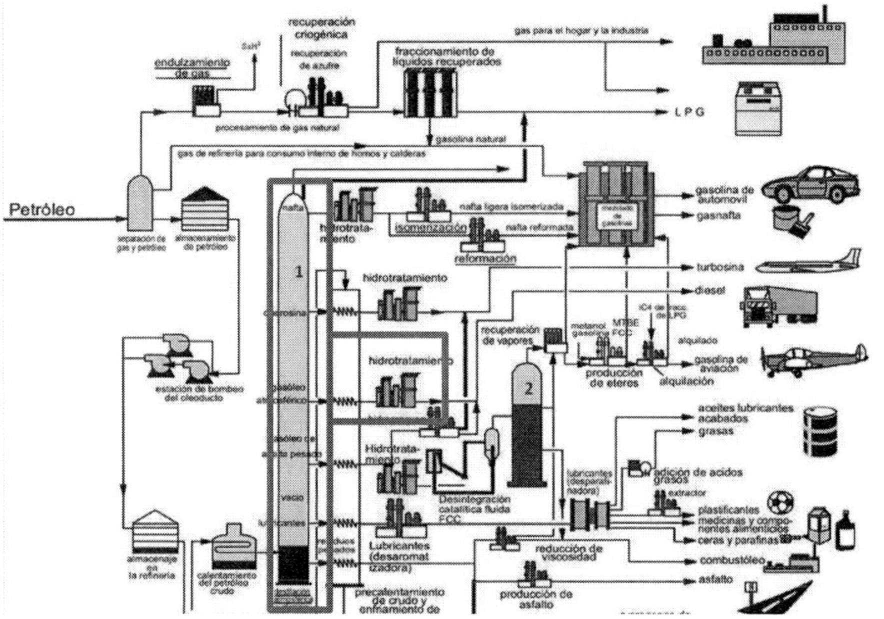

Chemical and Environmental Engineering Department
Universidad Nacional de Colombia

Environmental Engineering
Universidad de Cundinamarca

4

PROBLEM STATEMENT

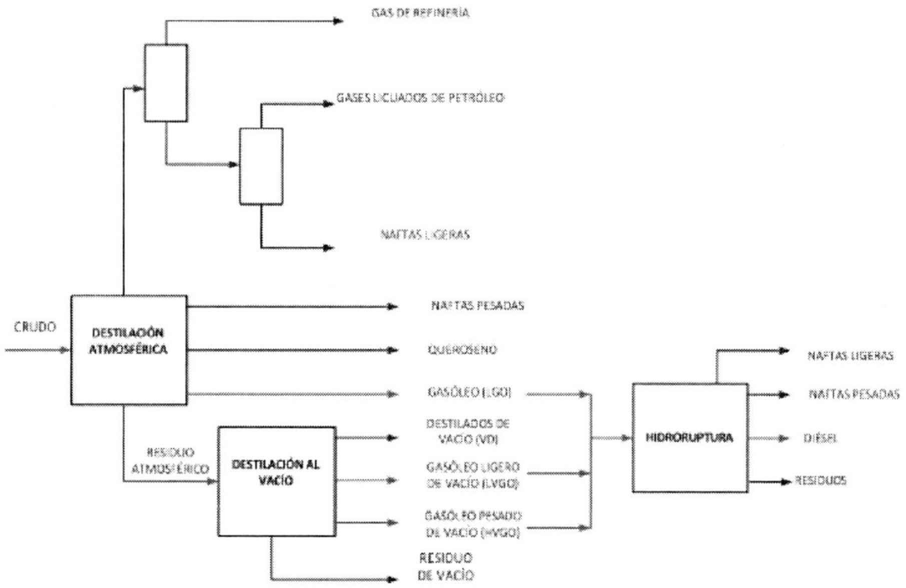

Chemical and Environmental Engineering Department
Universidad Nacional de Colombia

Environmental Engineering
Universidad de Cundinamarca

5

PROBLEM STATEMENT

OPTIMIZATION

- Reduce greenhouse gas emissions
- Decrease energy consumption
- Increase the Diesel produced
- Reduce operational costs
- Increase profitability

Chemical and Environmental Engineering Department
Universidad Nacional de Colombia

Environmental Engineering
Universidad de Cundinamarca

6

General Objective

Evaluate through meta-heuristic multiobjective optimization, the operational and processing intervals, in order to minimize the emission of greenhouse gases and maximize the profitability of the process.

Chemical and Environmental Engineering Department
Universidad Nacional de Colombia

Environmental Engineering
Universidad de Cundinamarca

7 METHODOLOGY

Aspen-HYSYS ⟷ MATLAB®

Modeling and Process Simulation

- Southeast Asian refinery conditions (validation of the simulation)
- Sensitivity analysis
- Evaluate the objective functions

Programming Bioinspired Algorithms

- Extract the results of Hysys
- Obtaining optimal pareto fronts

Chemical and Environmental Engineering Department
Universidad Nacional de Colombia

Environmental Engineering
Universidad de Cundinamarca

8 METHODOLOGY

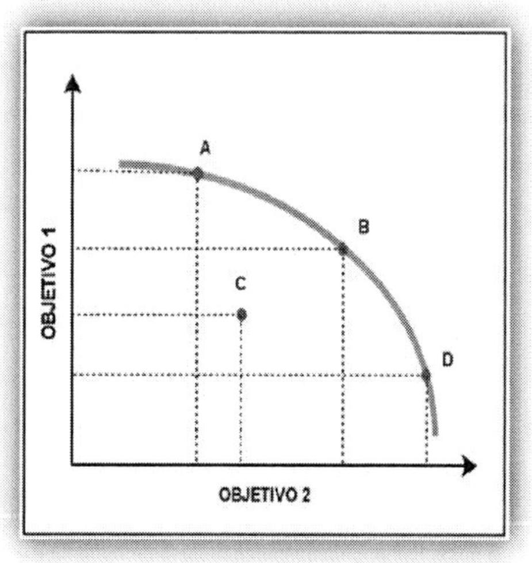

NSGA-II **m-NSPSO**

Chemical and Environmental Engineering Department
Universidad Nacional de Colombia

Environmental Engineering
Universidad de Cundinamarca

9 RESULTS

- SENSITIVITY ANALYSIS
- APPROACH TO THE OPTIMIZATION PROBLEM
- SELECTION OF THE OPTIMIZATION ALGORITHM
- OPTIMIZATION OF THE PROCESS

10 SIMULATION

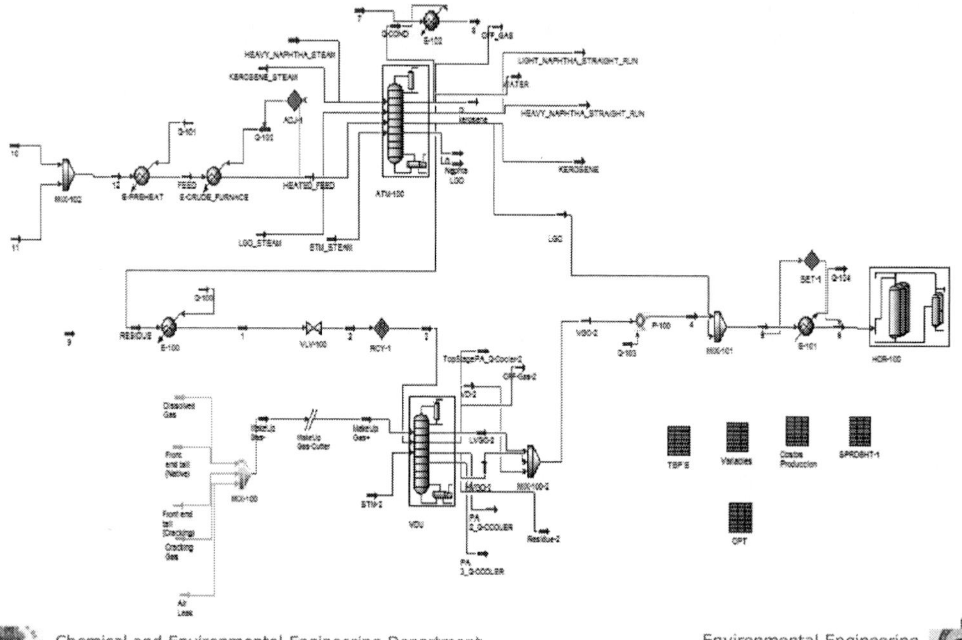

11

SENSITIVITY ANALYSIS

- 20 manipulated variables (independent variables)
- 30 output variables

$$\% \, Variation = \left[\frac{\max value - \min \text{value}}{\max value} * 100 \right]$$

Flow (kg/h)	Greenhouse effect gases					
	Feed HC	kerosene	LGO	LVGO	HVGO	VGO
Top gas	30,46	55,34	59,22	44,37	29,87	40,56
Light Naphta	68,64	72,73	72,23	70,54	48,01	59,50
Heavy Naphta	36,93	76,48	64,51	60,84	41,95	38,03
Kerosene	81,13	92,47	93,08	81,71	38,81	51,32
LGO	80,98	59,41	89,53	90,16	86,63	86,96

12

OPTIMIZATION PROBLEM

Profit function: The profit function is calculated as follows

INAMDAR et al 2014

$$\phi = P_1 - P_2 - P_3 - \text{Energy Cost}$$

where

ϕ = Profit (Rs./day)

P_1 = Selling price of the products (Rs./day)

$= \sum_i (\text{Prod}_i \times \text{Price}_i)$; i = 1, 2, ..., 7,

P_2 = Cost of raw materials (Rs./day)

$= (F \times \text{Cost}_f)$

P_3 = Operating cost (Rs./day)

$= \sum (RF \times RC + PA \times PC + Q \times UC)$

RF: Reflux flow rate (MT/day)
RC: Cost of pumping reflux (Rs./MT)
PA: Total pump around flow rate (MT/day)
PC: Cost of pumping PA flows (Rs./MT)
Q: Quantity of utility, e.g., steam rate (MT/day)
UC: Utility cost (Rs./MT)

13 OPTIMIZATION PROBLEM

- Max Profitability
- Min Greenhouse Gases.

- Subject to:

- Top gas flow [10000 - 40000 kg / h].
- Light naphtha flow [30000 - 150000 kg / h].
- Heavy naphtha flow [10000 to 50000 kg / h].
- Kerosene flow [10000 to 80000 kg / h].
- LGO flow [50000 to 200000 kg / h].
- Crude mix percentage [20% to 80%].
- Restriction in the quality of solvents

Chemical and Environmental Engineering Department
Universidad Nacional de Colombia

Environmental Engineering
Universidad de Cundinamarca

14 OPTIMIZATION PROBLEM

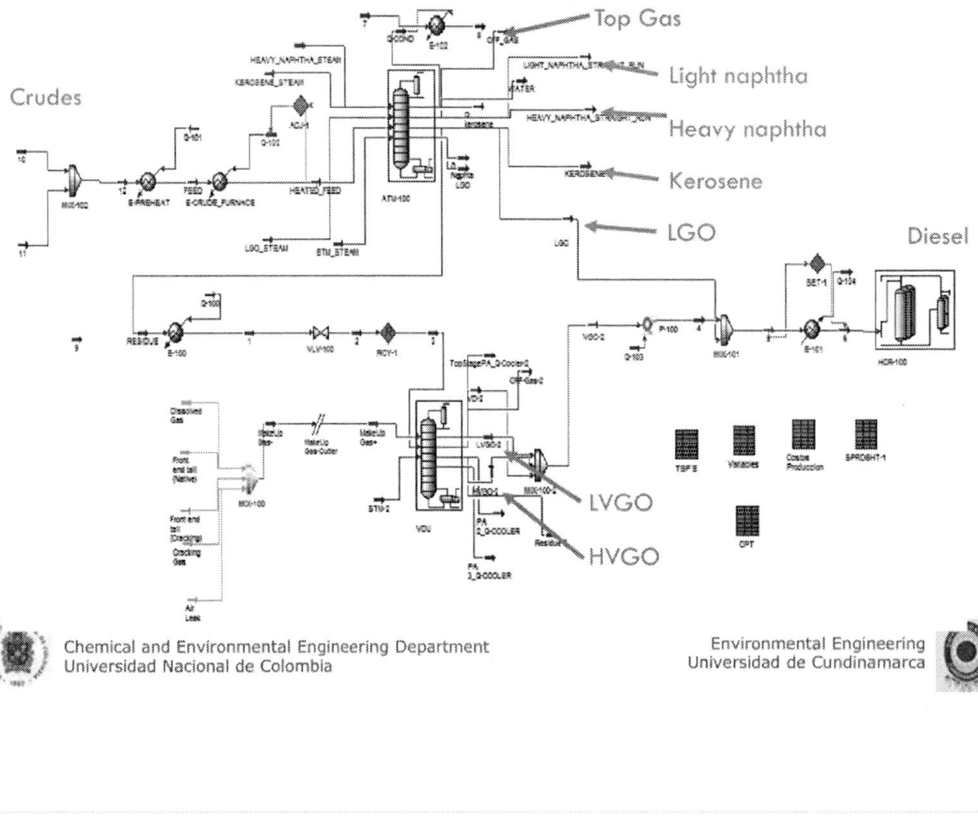

Chemical and Environmental Engineering Department
Universidad Nacional de Colombia

Environmental Engineering
Universidad de Cundinamarca

15

OPTIMIZATION PROBLEM

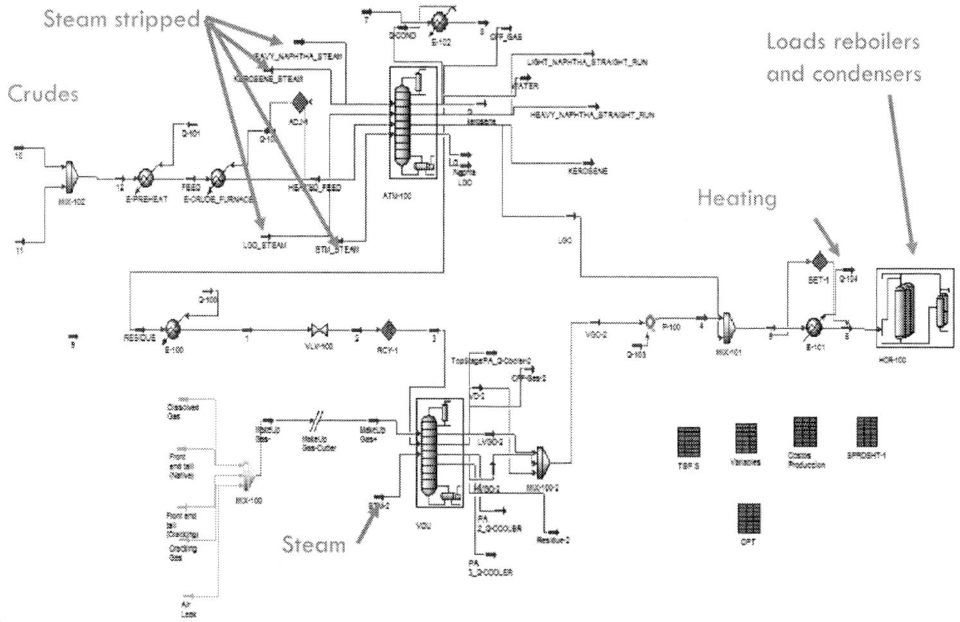

Chemical and Environmental Engineering Department
Universidad Nacional de Colombia

Environmental Engineering
Universidad de Cundinamarca

16

Mono-objective Test Functions

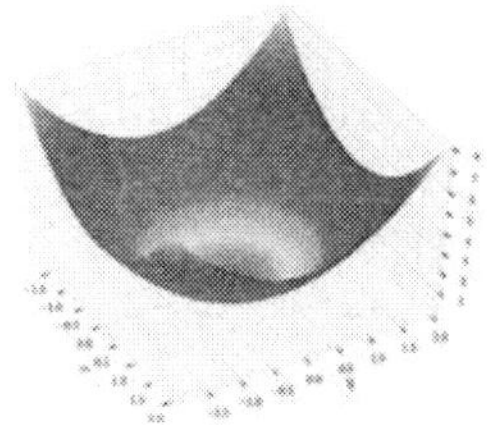
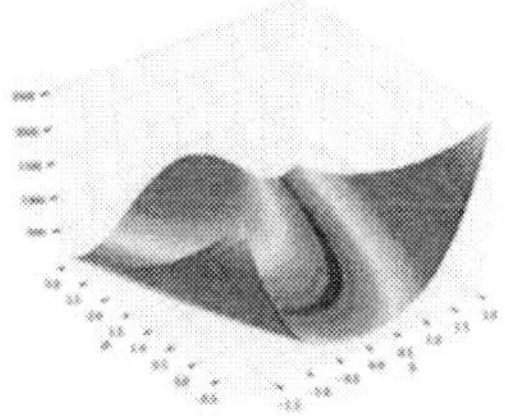
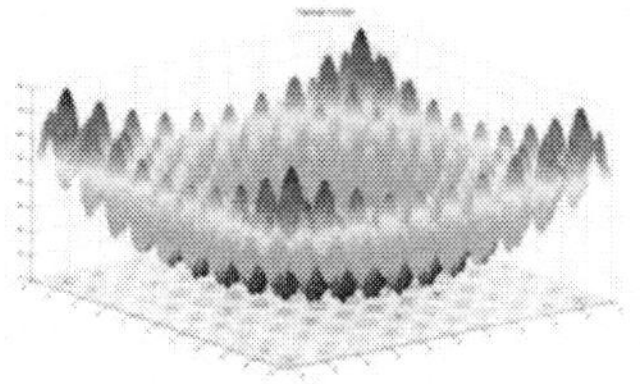

- Cuadratic Rosenbrock Rastrigin

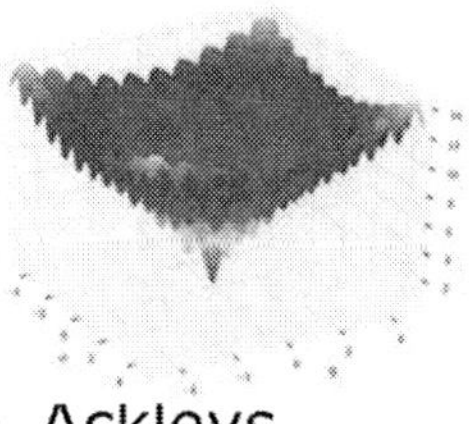
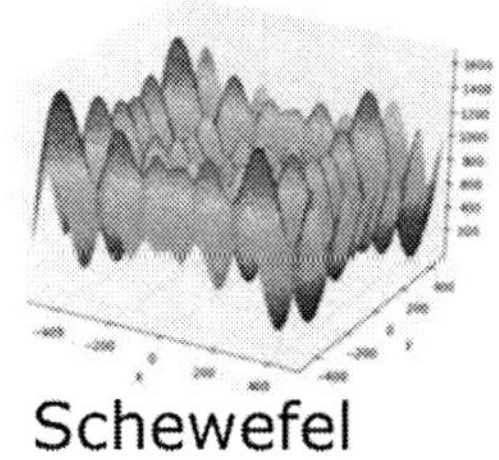
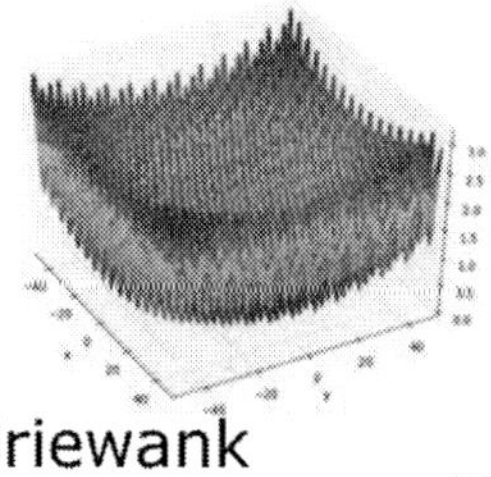

- Ackleys Schewefel Griewank

Chemical and Environmental Engineering Department
Universidad Nacional de Colombia

Environmental Engineering
Universidad de Cundinamarca

17

Mono-objective Test Functions

- KUR

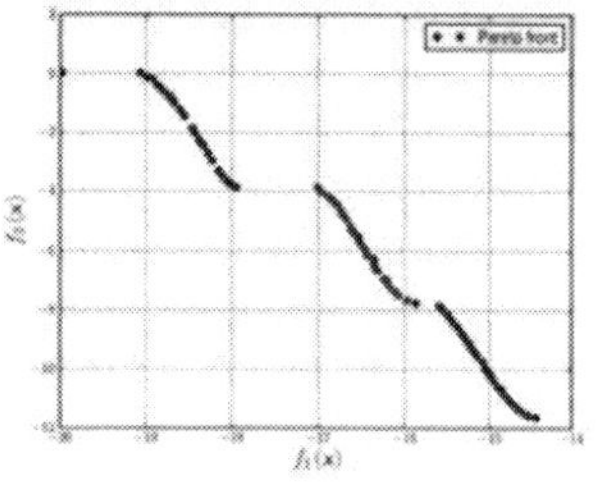

ZDT3

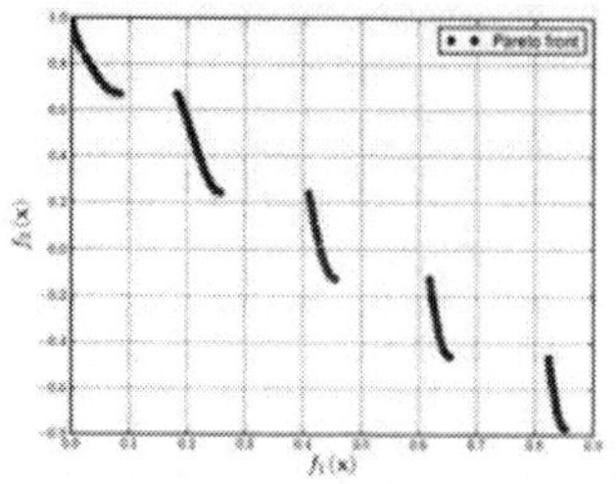

- ZDT4

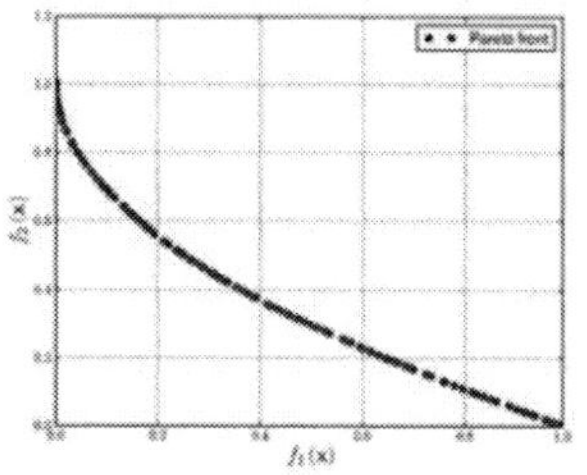

ZDT6

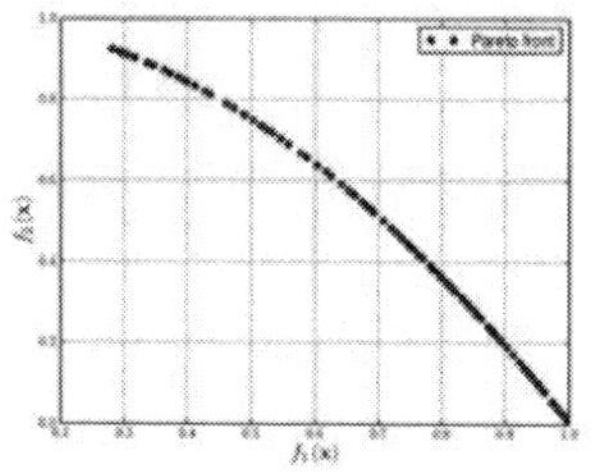

Chemical and Environmental Engineering Department
Universidad Nacional de Colombia

Environmental Engineering
Universidad de Cundinamarca

18

Algorithm selection

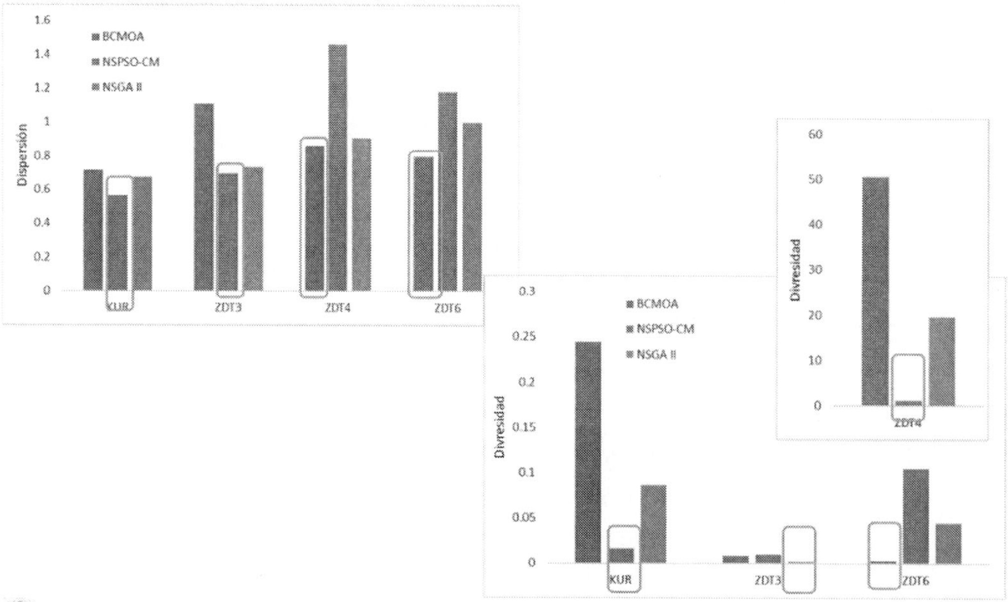

Chemical and Environmental Engineering Department
Universidad Nacional de Colombia

Environmental Engineering
Universidad de Cundinamarca

19

Algorithm selection

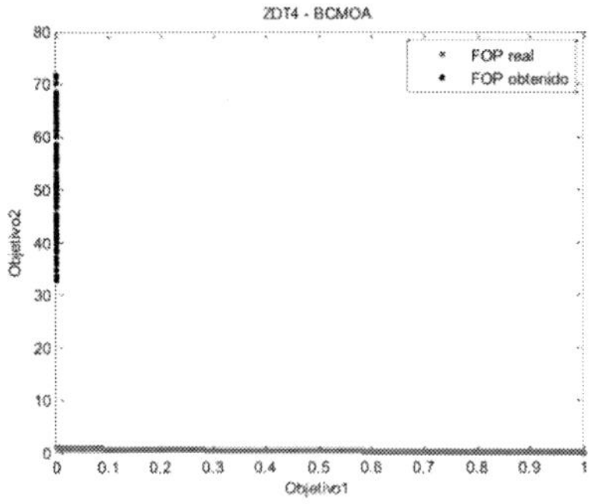

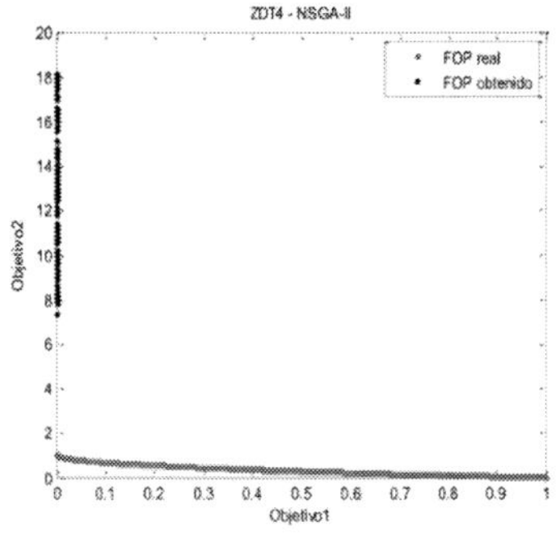

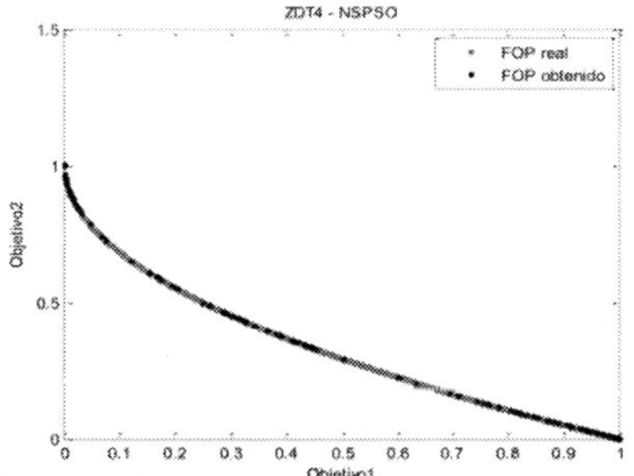

20

Results Optimization

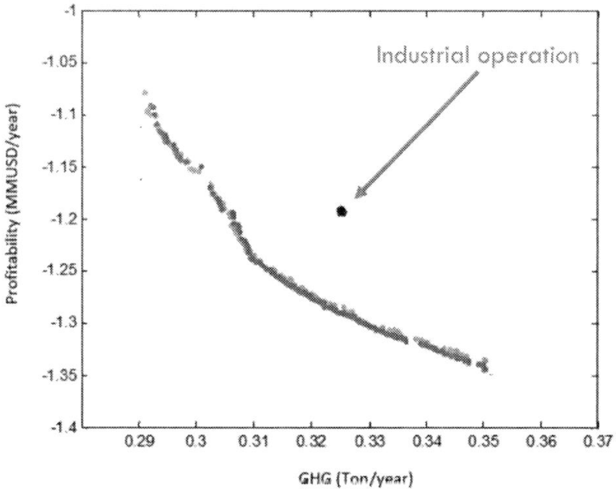

Figura 2. FOP obtained with the NSGA-II algorithm (red), NSPSO (green) and current point operation (black)

Chemical and Environmental Engineering Department
Universidad Nacional de Colombia
Environmental Engineering
Universidad de Cundinamarca

21 Resultados Optimización

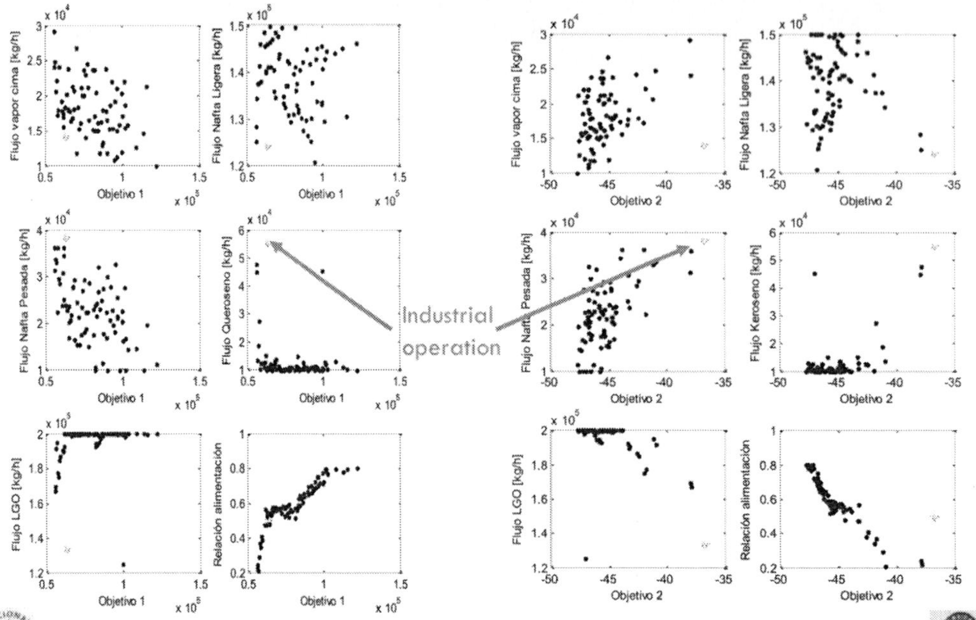

Chemical and Environmental Engineering Department
Universidad Nacional de Colombia

Environmental Engineering
Universidad de Cundinamarca

22 Conclusions

- It was found that the mNSPSO algorithm obtained the best performance
- We obtain a non-dominated set of optimal solutions that allow the maximization of profitability and the maximization of the emission of greenhouse gases.
- Two points of operation stand out from this set of optimal solutions: The first maximizes the quality of the products: peak steam flow (23900 kg / h), light naphtha flow (125130 kg / h), heavy naphtha flow (36089 kg) / h), kerosene flow (47475 kg / h), LGO flow (166960 kg / h) and percentage of crude mixture (22%). The second maximizes the flow produced by diesel: peak steam flow (10000 kg / h), light naphtha flow (146100 kg / h), heavy naphtha flow (10000 kg / h), kerosene flow (10000 kg / h) , flow of LGO (199730 kg / h) and percentage of crude mixture (80%). Both solutions improve the current ones in the refinery.

Chemical and Environmental Engineering Department
Universidad Nacional de Colombia

Environmental Engineering
Universidad de Cundinamarca

23

Chemical and Environmental Engineering Department
Universidad Nacional de Colombia

Environmental Engineering
Universidad de Cundinamarca

Development of a Computational Fluid Dynamic modelling concept for the RecoDust-Process

Franz Edler [a], Wofgang Reiter [a], Bernhard Geier [a], Johannes Rieger [a], Christoph Spijker [b], Harald Raupenstrauch [b]

[a] K1-MET GmbH, Linz, Austria
[b] Chair of Thermal Processing Technology, Montanuniversitaet Leoben, Leoben, Austria
Corresponding author: Franz Edler, Franz-Josef-Strasse 18, 8700 Leoben, Austria
E-mail: franz.edler@k1-met.com

Keywords
RecoDust, Flashreactor, CFD modelling, industrial burner, dust treatment, steel industry, atmosphere particle kinetic model, advanced Flamelet Equilibrium Hybrid model with discrete ordinate model, steel mill dust recycling, zinc reduction

Abstract
The **RecoDust process** offers new possibilities for minimum waste strategies in the iron and steel industry. The idea was realized in the Flash-Reactor pilot plant, whereas the **RecoDust process** represents an innovative pyrometallurgical treatment for steel mill residuals such as dusts with low zinc contents (~ 5 wt-% to 15 wt-%). In the framework of this project, the goal is to further develop the Flash-Reactor regarding the burner geometry, the dust feed charging system and the reactor design. This will be realized by means of Computational Fluid Dynamics (CFD) case studies whereas detailed simulation models are implemented. For a scale-up of the current pilot plant a mathematical CFD model is set-up.

RecoDust-Process

For the steel mill dusts with varying zinc contents a recycling process is limited or impossible without additional treatment. The **Waelz-Kiln** as state of the art treatment is only economical with zinc contents over 20 %. The dust from the blast furnace route possesses zinc contents between 5 wt-% and 15 wt-%, so they have to be recycled externally or landfilled. Aim of the innovative **RecoDust process** is the production of two fractions: A zinc rich oxide fraction and an iron rich fraction. The purity of the products has to be high enough to ensure that both fractions can be used as secondary resources in the zinc or iron industry. The process uses a reducing atmosphere provided by an oxygen/natural gas burner with temperatures around 1700 °C. Within these conditions, the fine dust melts in a reaction vessel and the zinc oxide is reduced and vaporized.

The zinc gas leaves the reaction chamber with the exhaust gas and oxidizes to zinc oxide dust in a post combustion tube. In a final step, the exhaust gas is cooled and subsequently zinc oxide is separated from the flue gas in a bag filter system. The nonvolatile components gathered on the bottom of the reaction vessel are discontinuously tapped as a slag. The benefit of the **RecoDust process** is the absence of foregoing treatment steps for the dusts.

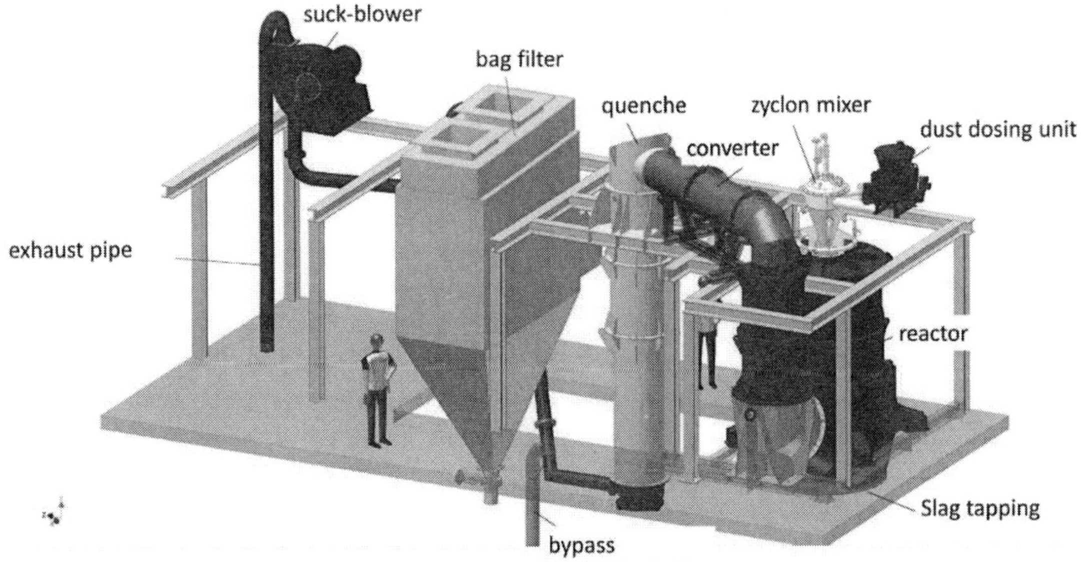

Figure 1: 3D-sketch of the Flash-Reactor pilot plant.

However, to ensure a fast melting procedure, a free floating and dry dust is limited to a grain size less than 1 mm. Unlike most other established processes there are no additives charged into the process, and the reducing conditions are only controlled by the oxygen flow (air excess ratio) in the mixing cyclone and the burner lance. The Flash-Reactor pilot plant as shown in Figure 1 installed in the technical center at the Chair of the Thermal Processing Technology at Montanuniversitaet Leoben has a dust mass throughput up to 300 kg/h.

Simulation concept

The direct-fired dust reactor is in the focus of the development with CFD. A sub-stoichiometric methane / oxygen flame with an air excess ratio < 1 is used in the burner. The main process parameters are particle size and composition, gas composition, particle and gas temperature and burner and reactor geometry. The simulation concept has two blocks, Block A and Block B.

Block A includes fast models to support the campaign planning and evaluating the results on the pilot plant. With the **"Atmosphere Particle Kinetic Model"**, it is possible to check the reactions in range of the calculation settings [1] as shown in Figure 2.

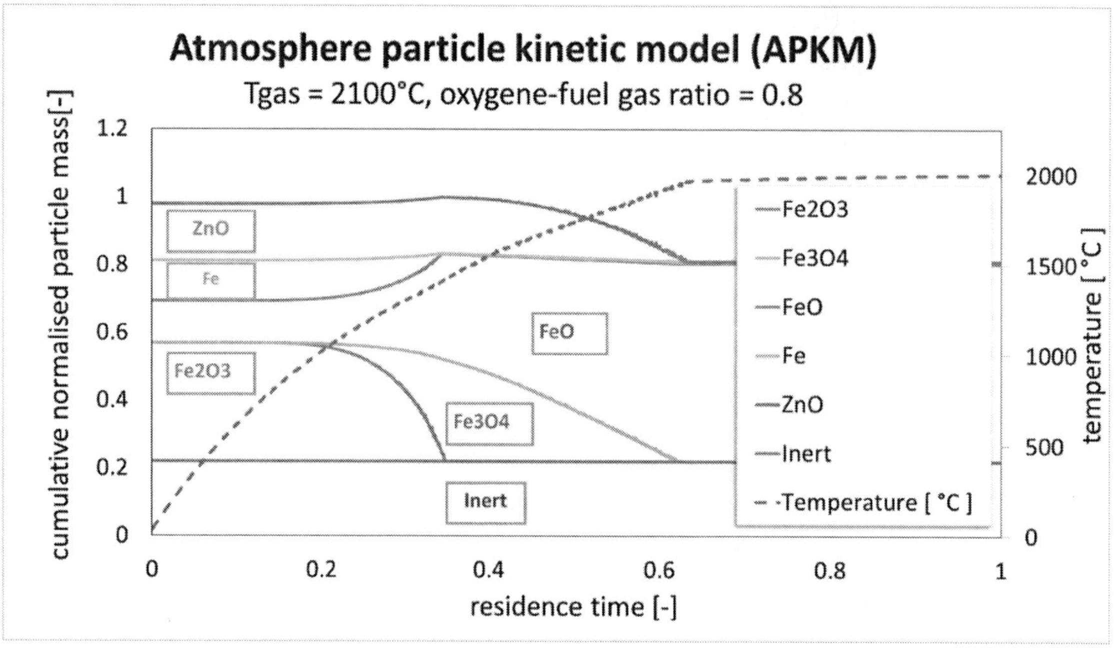

Figure 2: Results of the atmosphere particle kinetic model. Cumulate mass and particle temperature of the composition in dependence on the particle residence time.

The first test case was calculated by using the computational fluid dynamic program ANSYS Fluent v15. The fluid transport is solved in Eulerian framework and the dispersed phase is modeled by using Lagrangian approach. Turbulence model used in this work is realizable-k-ε model, which produces the most stable results. The flue gas composition is calculated with a reduced 17 species reaction mechanism [2] designed to model natural gas combustion. User defined function has been used for particle reactions and heat and mass transfer between fluid and dispersed phase. The mathematical description is based on an assumption of a non-porous dust particle with different compositions and size distributions.

In **Block B** of the simulation concept, the CFD program OpenFoam v2.4 [3] is used. For considering the chemical reaction kinetics, a Flamelet solver based on LibOpenSMOKE [4] [5] is implemented. An advanced Flamelet Equilibrium Hybrid model with discrete ordinate model (DOM) for the radiation [6] and transient thermo parcels, with higher stability of the calculation system are integrated. The flue gas composition is calculated with the above-mentioned reduced species mechanism [2]. The fluid transport is solved in an Eulerian framework and the dispersed phase with a Lagrangian approach. The realizable-k-ε model was used to describe turbulence phenomena [4].

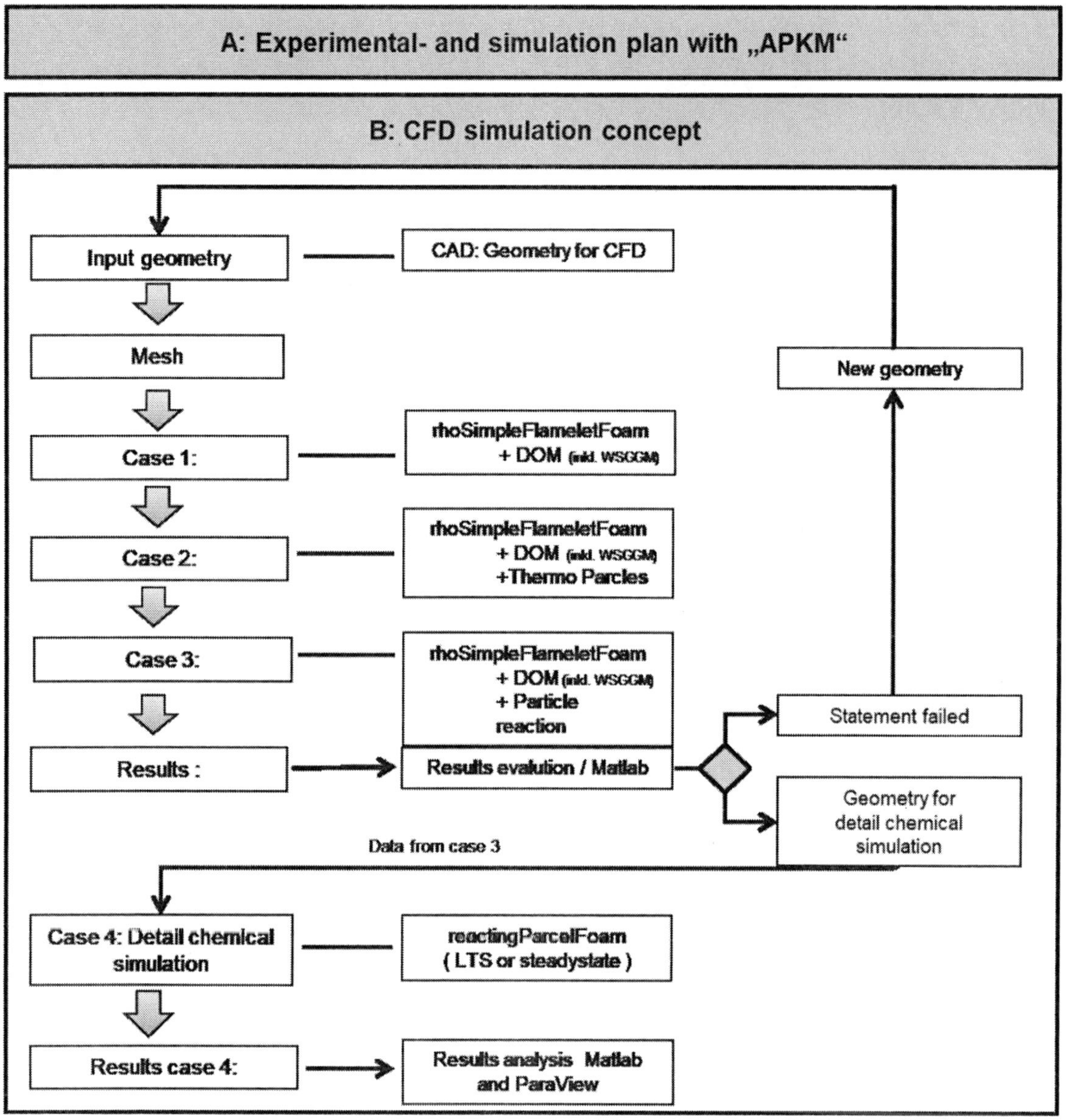

Figure 3: Scheme of Simulation concept.

Figure 3, shows a schematic view of the simulation concept, which describes the order of calculation results from Case 1 to Case 3 to analyze the influence of geometry on process efficiency. With the Flamelet Equilibrium Hybrid model including transient thermo parcel it is possible to check new burner designs and reactor designs in case of gas flow and particle flow through the Flash-Reactor. The velocity distribution, temperature distribution and the particle transport out of the reactor (= carry over) are important system parameters to evaluate the geometry design. In case that the assumed reactor design and process parameters meet the expectations in terms of process efficiency, next step of the simulation concept can be started (= Case 4). This case includes the calculation of the particle kinetic coupled with the fluid phase. The solution from the Flamelet Case 3 transferred to an OpenFOAM Partially Stirred Reactor model with transient parcel tracking. After the detailed chemical calculation, it is possible to evaluate the particle and the slag distribution from the wall film, the bottom slag or the particle carry over into the exhaust pipe, respectively. These results enable a comparison of calculated with experimental data.

Results and discussion

It is mandatory for the scale up of the pilot plant to understand the conditions inside the reactor. In Figure 4, the particle distribution colored by particle diameter in the reactor and in the pipe to the converter. The particles follow the velocity streamlines depending on the particle size and density. In Figure 4, the particles colored by diameter within the fluid volume are plotted.

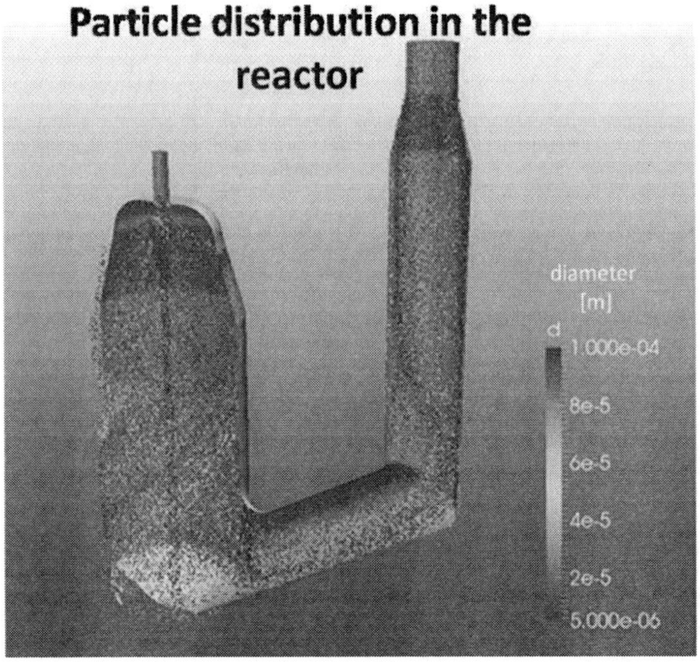

Figure 4: Particle distribution in the reactor

In the Figure 5, explicit chosen calculated results of the three-dimensional CFD case studies are shown. As discussed above, the fluid condition and the flame shape are important for the process. The experimental data quantifies the particle carry over out of the reactor is being influenced by the velocity distribution inside the reactor. The shape of the flame and high gradients from the temperature distribution, as well as the residence time are important for the reduction of the zinc oxide. As shown in Figure 5 below on the right, the velocity streamlines confirm the inhomogeneous conditions in the reactor. This means that due to the burner geometry and the resulting mixing behaviour inside the reactor, an intensive contact between solids and gas is realized leading to a high zinc conversion rate. In Figure 5 on the top right above, the input of zinc oxide for different particle classes in the wall film is shown. The particle size is in the range between 1 um (class 1) and 1000 um (class 10).

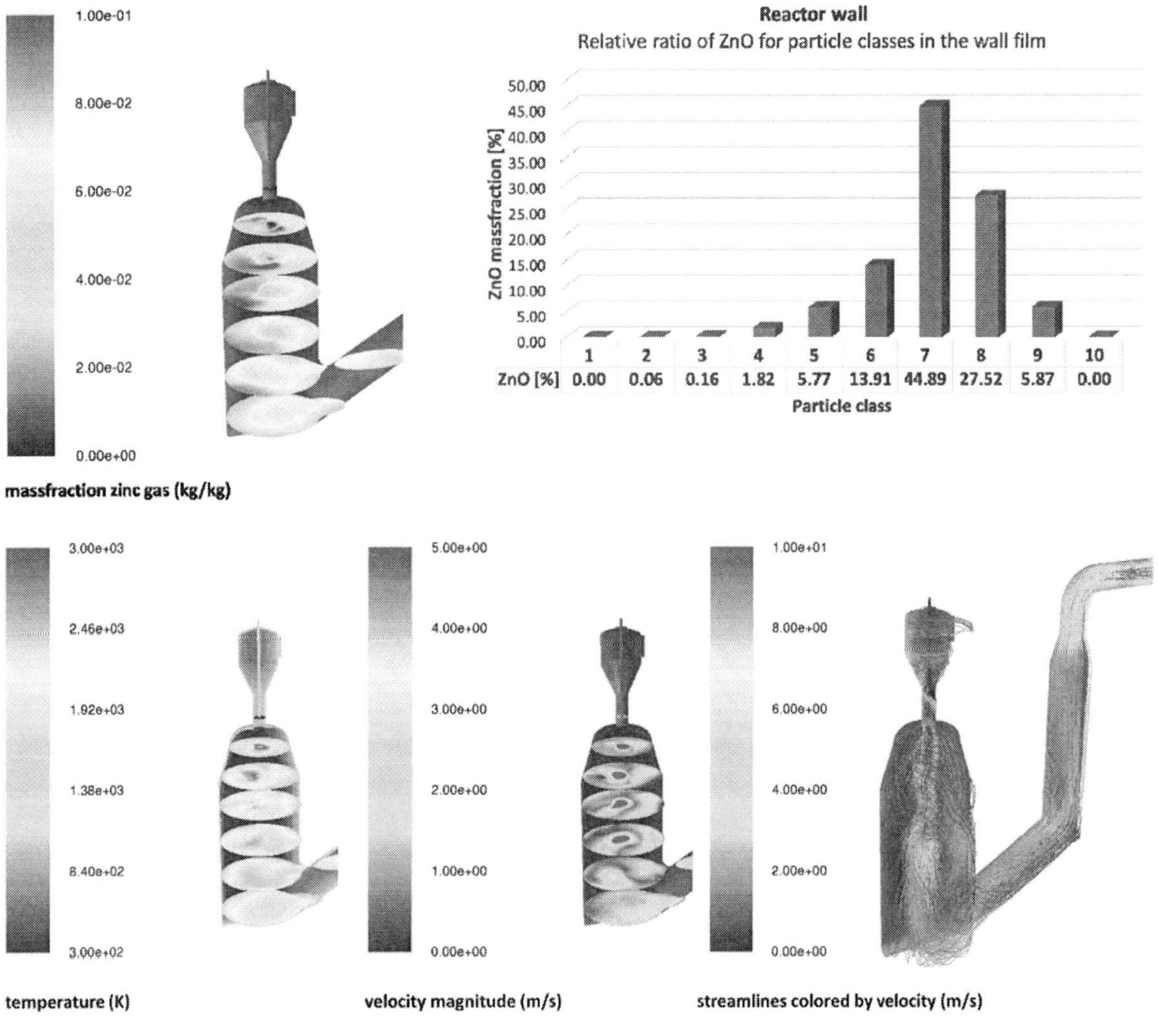

Figure 5: Results from the CFD calculation and Data evaluation with Matlab. Figure on the top left above: Mass fraction of zinc (g) in different cross sections of the reactor. Figure on the top right above: Data evaluation with MATLAB regarding the relative ratio of ZnO contained in each particle class at the reactor wall film. Figure below on the left: Temperature distribution in different cross section of the reactor. Figure below in the middle: Velocity magnitude in different cross sections of the reactor. Figure below on the right: Streamlines colored by velocity magnitude.

Outlook

The implementation of simple Flamelet models for case 1 to case 3 is completed. For future work, the particle reactions in the OpenFOAM "partial stirred reactor model solver" will be implemented. Finally, all Simulations are able to run on OpenFOAM and a routine for result analysis could be programmed.

Acknowledgements

The authors gratefully acknowledge the funding support of K1-MET GmbH, metallurgical competence center. The research program of the competence center K1-MET is supported by COMET (Competence Center for Excellent Technologies), the Austrian program for competence centers. COMET is funded by the Federal Ministry for Transport, Innovation and Technology, the Federal Ministry for Science, Research and Economy, the province of Upper Austria, Tyrol, and Styria and the Styrian Business Promotion Agency.

References

[1] F. Edler, "Development of an atmosphere particle kinetic model for particle reactions in a combustion Flash-Reactor using CFD- methods.," in *Energy Procedia*, 2017.

[2] "Cantera," [Online]. Available: http://www.cantera.org. [Accessed 11 04 2017].

[3] S. Thynell, "Discrete-ordinates method in radiative heat transfer," *International Journal of Engineering Science Volume 36,* pp. Issues 12-14, pp. 1651-1675, 1998.

[4] C. Spijker, "Unsteady Laminar Flamelet Modellierung zur Beschreibung von Mündungsmischbrennern," Montan Universitaet Leoben, 2010.

[5] "OpenFOAM," [Online]. Available: http://www.openfoam.com/. [Accessed 11 04 2017].

[6] T. Holzmann, "www.holzmann-cfd.de," [Online]. Available: http://www.holzmann-cfd.de/index.php/de/flamelet-model/14-development. [Accessed 11 04 2017].

[7] A. Couci, A. Frassoldati , T. Faravelli und E. Ranzi, „libOpenSMOKE, User-Guide," 2012.

499467 Dynamic Simulation and Optimization for Gas Expanded Liquid Phase of Ethylene Oxide Production with Free Carbon Dioxide Emission

Abstract:

Ethylene oxide production is considered as one of the processes that produce carbon dioxide as byproduct. In response to the global warming issue, the following work illustrates alternative synthesis to produce ethylene oxide with free carbon dioxide emission. The new procedure was found by The Center for Environmentally Beneficial Catalysis (CEBC), University of Kansan and improved by Laboratory of Integrated Systems Engineering (LISE), Lamar University. The ethylene oxide is produced by reacting ethylene and hydrogen peroxide which is the reaction oxidant over methyltrioxorhenium catalyst (MTO) in a continuous stirred reactor at 104 áµ'F and 725.18 psia.

At LISE, the new steady state design has been built on Aspen Plus version 8.8, focusing on solving some of the problems that have been mentioned by CEBE and improving the distillation and purification process. Therefore, the design is consisted of two continuous stirred reactors, a flash drum, a trays column, and two packed bed columns.Â In response to the waste of unreacted hydrogen peroxide, a better design layout set has been developed to recycle and regenerate that hydrogen peroxide back to the main epoxidation reactor. Fortunately, these adjustments lead to optimize the alternative ethylene oxide synthesis, it decreases the production cost to 52 ¢/lb. of ethylene oxide, and the payback period for new ethylene oxide production plant is 3.48 year which is evaluated by Aspen Process Economic Analyzer.

For further studies on the process safety and process operation stability, a dynamic well controlled model is built in Aspen Dynamic version 8.8. The safety of the process has been assigned as the first priority, and the process capability is the second priority in designing the control scheme and control parameters tuning for the dynamic model. First, the conversion in the main epoxidation reactor is controlled by the reactor temperature to specified value, so the catalyst stays as active as possible. Second, with the hydrogen peroxide decomposer, the decomposition reaction is controlled by the reactor temperature to ensure that all the hydrogen peroxide is decomposed before entering the separation and purification step. As well as, the reactor vapor outlet composition is controlled by the reactor pressure to keep that out of the flammability zone for ethylene oxide, ethylene, and methanol (which is reaction solvent). Third, with the hydrogen peroxide recycle stream, the decomposed oxygen is regenerated to hydrogen peroxide; therefore, the main control objective is to make sure all oxygen is converted to hydrogen peroxide before they reenter the epoxidation reactor. Fourth, with the distillation columns, the main objective for the first tower is to recycle most of the unreacted ethylene back to the epoxidation reactor. On the second column, the objective is to get the water in the bottom stream to be pure as much as 99.995% mol. On the last column, the objective is to achieve the desired market purity for ethylene oxide. Finally, by applying all the previous objectives and specifications to the model, the process operation stability could respond on any disturbance that is generated by upstream set point changes up to 15% of its steady state values.

Also, we have gone further, and we rebuilt the control system in the DeltaV DCS system to control the simulated process through OPC Client server, so we can have a more realistic and reliable model to the real plant operation. Right now, we are working on alarm rationalization study using DeltaV system, and it will be done by the conference time. Beside that work, we will make a model predictive control model which will be generated by Aspen DMC Plus and DeltaV that will be used to do further study on dynamic optimization and real time optimization.

488800 Isolation of Low Grade Biobased Bulk Chemicals

Abstract:

In many applications present biorefinery suffers from the problem of limited utilization of products formed. Pulping of wood produces about or even less than 50% cellulose, while several byproducts need to be incinerated because of economic reasons. Typically, black liquor from pulping does contain appreciable amounts of carboxylic acids besides sugars and other byproducts. Targeting complete utilization of these low grade bulk chemicals does need severe technological intensification of downstream processing.

Focus of this project was the intensification of isolation of acetic acid and formic acid from black liquor evaporation condensates. This ternary mixture exhibits a binary high boiling azeotrope and a ternary saddle point azeotrope with the low boiling component water in high excess. Liquid-liquid extraction is well investigated and perfectly suited for separation of the carboxylic acids from the aqueous feed. However, solvent regeneration and byproduct isolation are still too energy intensive.

Combining distillation with an appropriate chemical reaction may intensify the solvent regeneration. Esterification of the carboxylic acids with a low molecular weight alcohol produces low boiling esters. These esters can be continuously removed from the reaction broth by distillation, acting positively on the equilibrium composition of the aqueous broth towards complete conversion of the target constituents. Expectedly catalytic acceleration of esterification would contribute to intensification. By using a reactive extractant with preference to acids the liquid-liquid equilibrium is improved compared to physical extraction and the acidic catalyst is kept in the solvent phase during extraction due to adduct formation with the reactive extractant. Admixture of solvent-soluble acidic coextractants may for sure limit the extraction capacity of the solvent, but provide the catalytic support. In solvent regeneration via reactive distillation the elevated temperature in the reboiler releases the catalyst from the adduct.

For technology assessment and process validation experiments were performed with Cyanex®923 (a Lewis basic phosphonate extractant) for extraction plus 4-dodecylbenzenesulfonic acid for catalysis. With this solvent mixture complete conversion and isolation of formic acid was achieved and methyl acetate yield of 88 % was obtained. For comparison distillative isolation of acids would cause an economic disaster and isolation by extraction/distillation would end up in the ternary azeotropic mixture.

498255 An Efficient Approach for Chemical Process Development Using Kinetic Modeling in Batch and Continuous Mode

Abstract:

Traditionally active pharmaceutical ingredients are manufactured in batch mode, however the pharma companies express increasing interest in continuous processing. Continuous manufacturing is an attractive alternative due to its benefits such as reduced development time, greater process safety when employing hazardous chemistries or extreme conditions, reduced operating costs, improved process control and product quality.

Kinetic modeling plays an important role in development stages of chemical processes regardless of the operation mode. Precise knowledge of the kinetic parameters is necessary for the scale up from laboratory to industrial scale and for the design of robust and safe processes.

This work describes an efficient approach for the development of a reaction in an API synthesis. The approach includes screening of the reaction conditions, estimation of the kinetic parameters both in batch and flow mode and their comparison, and also fine-tuning optimization applying Design of Experiments.

In the screening phase, we tested both conventional and microwave heating to find the most favorable conditions resulting in 99% conversion and we analyzed the possibilities for a continuous process. Specific sets of experiments were performed in both operation modes, kinetic curves and parameters were then obtained and assessed to determine if the studied variables impacted the impurity formation. The model was used to predict conversion values and impurity levels for different conditions. In the last stage optimization studies were carried out in order to determine the Design Space ensuring the product quality within specifications.

We demonstrate our methodology applied for a case study following the Quality by Design concept, including experimental data and also simulations. This approach can be used for the development of batch or continuous processes in the future. The use of kinetic modeling allowed deeper understanding of the reaction mechanism and its influential factors, resulting in shorter development time and enhanced confidence level regarding the impurity formation.

487968 Intensification of Carboxylic Acid Isolation Processes By Emulsification

Abstract:

Process intensification holds a central role in the biorefinery, as the substitution of crude oil by renewable feedstock overstrains standard processes in terms of economy and efficiency. Intensification can be achieved by applying heuristic methods to develop innovative processes in biomass refining. While conversion of biomass to any desired extend is the lesser problem (at least from the point view of downstream processing), the downstream processing of effluents is energy and resource demanding. Therefore, low grade carboxylic acids or low grade alcohols are rarely processed. Within isolation, low concentration and unfavorable combinations of separation relevant substance properties form the most troublesome problems. A promising concept to deal with this economic problem is the change of properties like solubility or relative volatility by chemical conversion.

The change of properties, a tool of process intensification, can be combined within existing processes and equipment. Esterification of low grade carboxylic acids for example does change physical properties significantly. For esterification of carboxylic acids (e.g. acetic acid or formic acid) high excess of water is a major obstacle. Superimposition of the chemical conversion with liquid-liquid extraction can be used to overcome this limitation. Enhancement of the reaction rate as well as mass transfer, may be attained by catalyst-induced emulsification. Surface active, acidic catalysts like sulfonic acids (e.g. *4*-dodecylbenzenesulfonic acid) accord with these requirements.

As a test system 60 g L^{-1} acetic acid was chosen with *n*-octanol as reactant and *n*-undecane as diluent. *4*-dodecylbenzenesulfonic acid was used as catalyst, due to its good performance in esterification catalysis and due to its emulsification properties. Experiments were performed in order to determine the influence of catalyst, reaction temperature and solvent composition on the reaction kinetics and separation efficiency. Experimental results for esterification of dilute acetic acid with *n*-octanol in the emulsion regime have shown the applicability of pseudo-homogeneous rate laws for modeling the reaction kinetics. This implies a quasi-homogeneous state as a result of emulsification. For implementation of this intensified process, a mixer-settler setup is suitable as it allows higher conversions and separation efficiencies through multi-stage operation. Phase separation can be achieved through centrifugation or electrical splitting.

487339 Sustainable Pig Iron Production Via Reduction of Mineral Iron Carbonate with Hydrogen

Abstract:

Innovative technologies are necessary to limit CO_2-emissions from energy-intensive iron and steel production from carbonate based ores. Iron carbonate reserves are found worldwide including reservoirs in the United States (e.g. Bisbee and Antler Mine in Arizona) and China. Pig iron is produced from iron ores in a multi-step process. This process includes roasting and sintering (Eq. 1) of iron carbonate minerals. Instead of roasting in air to produce hematite suitable for iron production via the blast furnace process, we suggest direct reduction of iron carbonate [1] with sustainably produced hydrogen (Eq. 2) as a 'green' alternative.

$FeCO_3 + 0.25\ O_2 = 0.5\ Fe_2O_3 + CO_2$ (1)

$FeCO_3 + H_2 = Fe + H_2O + CO_2$ (2)

Carbon dioxide emission (due to the thermal decomposition of iron carbonate according to Eq. 2) could even be diminished in a two-step process. In the first step, iron carbonate is directly reduced to iron and carbon dioxide. In the second step, the process gas is fed to a reactor for catalytic carbon dioxide hydrogenation to yield value-added carbonaceous products. In order to develop an environmentally benign synthesis route, an insight into reaction mechanism and kinetics is important. Therefore, thermogravimetric linear heating rate data was recorded and analyzed according to the ICTAC recommendations [2]. Iron formation according to Eq. 2 follows a nucleation and growth mechanism, verified by the prediction of the measured data with an Avrami-Erofeev A2 reaction model. Thermodynamic analysis revealed that the energy demand for iron production can partially be provided by the outlined process gas treatment by means of catalytic carbon dioxide hydrogenation to produce hydrocarbons.

1. Baldauf-Sommerbauer, G., Lux, S., and Siebenhofer, M. (2016) Sustainable iron production from mineral iron carbonate and hydrogen. *Green Chem.*, **18** (23), 6255–6265.

2. Vyazovkin, S., Burnham, A.K., Criado, J.M., et al. (2011) ICTAC Kinetics Committee recommendations for performing kinetic computations on thermal analysis data. *Thermochim. Acta*, **520** (1–2), 1–19.

495880 a Novel Selective Ion Separation Process Using Battery System

Abstract:

The increase of demand for rare metals and the decrease of the available ores make it interesting to develop separation processes for extracting metal ions from solutions such as brine lakes, seawater, and industrial waste water. In general, that solutions contain a mixture of salts and the concentration of valuable resource is very low, so it is required to develop a highly selective resource recovery technology. Distillation, adsorption and membrane processes are well-known separation technologies for ion separation from solutions, however, the selectivity, energy requirement, and operation costs are not sufficient. Therefore, a new approach to develop an efficient ion separation process is needed. In this work, we investigated the highly selective and energy efficient resource recovery process using battery system. The system consists of two battery electrodes which have ion selectivity. It operates a similar way to rechargeable batteries, and target ions are selectively extracted and released by charging/discharging step. Using this battery system, lithium or sodium ion are successfully separated from salt water even the solutions contain highly concentrated various ions (Yoon *et al., Phys. Chem. Chem. Phys.*, 2013; Yoon *et al., Energy Environ. Sci.*, 2014, Yoon *et al., Chemosphere*, 2015). This battery system has not only energy saving process with efficient energy consumption but also eco-friendly process without chemical adding during the operation. The battery system could be a suitable method for ion separation process for selectively extracting metal ions.

Reference

Jaehan Lee, Seung-Ho Yu, Choonsoo Kim, Yung-Eun Sung, **Jeyong Yoon**, Highly selective lithium recovery from brine using a Î»-MnO_2-Ag battery, *Physical Chemistry Chemical Physics*, 2013, 15, 7690-7695.

Jaehan Lee, Seoni Kim, Choonsoo Kim, **Jeyong Yoon**, Hybrid capacitive deionization to enhance the desalination performance of capacitive techniques, *Energy & Environmental Science*, 2014, 7, 3683-3689.

Seoni Kim, Jaehan Lee, Jin Soo Kang, Kyusik Jo, Seonghwan Kim, Yung-Eun Sung, **Jeyong Yoon**, Lithium recovery from brine using a Î»-MnO_2/activated carbon hybrid supercapacitor system, *Chemosphere*, 2015, 125, 50-56.

Â

496638 Computational Investigation of Liquid Maldistribution in Periodically Operated Structured Packed Beds

Abstract:

The non-linear hydrodynamics combined with multiscale transport processes engenders the modeling of packed bed reactors (PBR) a challenging task, particularly in addressing the scale up issues for its commercial application. Familiarized with the steady state hydrodynamics over the past two decades, researchers are now extensively exploring the efficacy of unsteady state operation of packed bed reactors in process intensification and prolong reactor service life. Despite the proven advantages, implementation of periodic flow operation in commercial PBRs is still far from reality, mainly due to process control safety. Although, the application of steady state PBR for commercial purposes is well established, the design concept of periodic PBR needs more attention, even heuristically, to understand the influence of all key parameters. With enhanced capabilities of CFD methods, computational studies can not only complement various facets of physical phenomena but also can help in understanding certain aspects that are generally not attainable by experiments. Surprisingly, limited researches have attempted to develop CFD based models for unsteady PBRs to understand the influence of operating parameters. Moreover, none of these studies have addressed the effect of packing arrangements on flow distribution during periodic operation. Additionally, hydrodynamics of periodic PBRs, fed with non-Newtonian liquids, has rarely been addressed. Cyclic or periodic operation, dictates the periodic toggling of a fluid phase at the inlet, between a low-level (base) and a high-level (peak), typically referred to as the min-max flow operation. This further is termed as the on-off flow when the base flow rate is set to zero.

In this article, we aim to develop a 3-D CFD model for periodic PBRs to understand the effect of particle arrangements and various splits under min-max and on-off operation on flow distribution. A CFD model is developed based on unit cell approach incorporating single phase periodic flow to understand the intrinsic behavior of flow modulation in structured packings. To elucidate the effect of non-Newtonian liquid, a basic power-law model is employed in our validated CFD model. To examine the effect of particle arrangements, face centered cubic (FCC) and modified simple cubic (SC) packing orientations are investigated as shown in Fig. 1. Figure 1: Schematic showing grids and boundary conditions of (a) FCC, and (b) modified SC arrangements. Moreover, three different combinations of min-max and on-off operation are chosen under fast mode to identify the influence of flow splits. Liquid distribution is quantitatively examined at the mid-plane of the geometries by dividing the plane into twelve area segments and then evaluating the mass flux through each of the segments. An improved distribution is observed for periodic flow as compared to continuous operation, which is even more pronounced for min-max operation. A distribution coefficient, calculated using the standard deviation concept, indicates that the FCC packing shows the best distribution for min-max operation with the split having maximum peak flow duration as shown in Fig. 2. However, exactly opposite trend is observed in modified SC, where the split with the lowest peak flow rate duration shows the best distribution among the three splits which was even more pronounced for the min-max operation. FCC exhibits better liquid distribution than the modified SC although having a higher pressure drop value due to its lower porosity. Consequently, a comparitive analysis was conducted to determine the desired mode of periodic operation with optimum split ratio that results in least maldistribution. This study will not only find relevance to several packed bed applications but also indicate the periodic flow operation as a viable process intensification technique.

Figure 2: Mass flux distribution at the mid-plane of FCC for different n values at t=10 s for (a) n=0.6, (b) n=1.0, and (c) n=1.2.

490671 Energy Saving from Process Design - a Service Oriented Architecture (SOA) Methodology

Abstract:

Energy saving is important and urgently needed in process manufacturing, which is energy intensive and takes ~75% of the total industrial energy use in US. To do the work, chemical engineers should take the lead role and process development & design has the greatest potential to achieve the goal.

However, it is very challenge to develop an optimal process design due to the complexity of the process. In general, there are four major challenges. First, the engineers or designers should have systematic knowledge and skills in process design. Second, we need to have an in-depth understanding of the whole studied process. Third, a successful process design implementation needs collaboration among designers, engineers, manufacturers, and operators. Fourth, a typical process design is a long-term development with dedicated specialists, which many companies especially small and medium sized companies could not afford.

To address the challenges, the author proposes a so called service oriented architecture (SOA) methodology. It is a framework to define the design scope and facilitate the work. Four case studies are presented which are based on the authorâ€™s industrial energy saving consulting work. It shows how the SOA methodology is applied into the real world and achieve accountable energy saving.

496499 Mathematical Modeling and Optimization of Commercial-Scale Catalytic Two-Stage, Ebullated Bed, Direct Coal Liquefaction Reactors

Abstract:

Due to the insecurity in crude oil supply and uncertainty in its prices, finding alternative fuel source is critical. Direct coal liquefaction (DCL) is one of the technically feasible solutions to convert coal to transportation fuels. However, this technology has not been widely commercialized due to the high capital investment, uncertainty in economic feasibility and several operating challenges. In the DCL technology, catalytic two-stage liquefaction (CTSL) unit with a configuration similar to the hydroprocessing reactor systems uses two ebullated bed reactors (EBR), in which coal reacts with the recycled solvent and hydrogen to produce syncrude. Since the conversion, and yield obtained in the CTSL reactors are probably the most important factor in the profitability of the DCL plants, the CTSL process must be optimized. A rigorous mathematical model of the CSTL unit is required for optimization. However, most of the models of the DCL reactors have been developed by assuming ideal reactor performance, with very simple reaction kinetics or by using simplified yield correlations developed using experimental observations.

In this work, a first-principles model of the CTSL reactors have been developed and validated by comparing with the commercial operating data. The ebullated-bed DCL reactors are simulated as axially dispersed reactor (ADR) with recycle stream based on rigorous reaction kinetics, hydrodynamics and mass and heat balances. Other standard unit operations, including heaters, pumps and flash drums, are modeled using mass and energy conservations. The capital cost of the reactors is estimated based on the pilot plant data, while that of other equipment items is estimated using Aspen Process Economic Analyzer (APEA). The CTSL unit is optimally designed to minimize an annuitized operating cost that includes operating and capital costs along with the revenue earned by selling the products. To summarize, the presentation will include the following: (1) two configurations of the CTSL unit- with and without heat integration, (2) rigorous modeling of the EBRs, (3) systems level models of the CTSL unit, (4) capital and operating cost estimates, (5) sensitivity studies of the key operating and design parameters, and (6) process optimization.

497031 Optimal Design of Pipeline System for Flare Minimization in Multiple Chemical Plants

Abstract:

Flare minimization (FM) in large-scale industry takes increasingly engineersâ€™ concerns nowadays. It is a win-win practice for not only environmental sustainability but also industrial profitability. A lot of researches have already focused on FM problems with many FM strategies for a single plant. However, there are few FM works focusing on multiple plantsâ€™ FM issues. Material exchanges (ME) among plants provide a new degree of freedom for material savings and flare reductions. For example, a start-up plant can borrow materials from another normal-working plant to help the plant compressor start-up, or a normal-working plant can receive off-spec materials from another plant undergoing the shutdown operation to help recover potential products. Such ME operations could be designed and scheduled based on plantsâ€™ operational flexibilities. Certainly, different plants may have different operating statuses when interplant ME are considered. The design of a pipeline system that satisfies different plant operating statuses is a must.

This work developed a general methodology for the optimal pipeline system design for FMs via interplant ME, which can be used for FMs under planed turnaround and emergency turnaround operations in multiple chemical plants. The operational scheduling together with the pipeline system design fully take interplant ME advantages while all possible plant operating scenarios have been considered. The methodology will provide a conceptual pipeline system design with the minimum capital cost, containing the information of detailed pipeline distributions and connections; control valve placements, and valve switching strategies under different interplant ME strategies for FMs. The efficacy of the developed method will be demonstrated by two case studies.

500436 Development of a Natural Gas to Industrial Chemicals Bioprocess Platform

Abstract:

The low cost and abundant supply of natural gas have been motivators for the development of many technologies that convert this feedstock into compounds of greater value. Intrexon has developed the first natural gas-to-liquids bioconversion platform powered by methantrophic fermentation, which applies synthetic biology to program bacteria to produce higher value materials of interest. Our innovative genetic toolbox allows us to engineer the biology of the methanotroph to more efficiently convert natural gas into a wide range of valuable chemicals and fuels. These applications include isobutanol, farnesene, 1,4-butanediol, 2,3-butanediol, isoprene and isobutyraldehyde – the building blocks for industrial and consumer products like fuels, synthetic rubber, acrylics, resins and spandex.

Using a methanotrophic bioprocess to produce fuels and chemicals has several advantages beyond the low cost and minimal processing of feedstock. Natural gas bioconversion offers scaling opportunities beyond traditional gas-to-liquid (GTL) facilities, as well as biomass feedstock facilities. Theoretical yields using the methanotroph and methane exceed that of yeast/sugar approaches by a wide margin. Biocatalytic processes that achieve high yields under milder processing conditions allow for reduced capital and operation expenditures, offering greater economic viability. The natural gas bioconversion also does not compete with food resources, unlike traditional biofuels that use sugar and corn as feedstock.

One molecule of interest in the methanotroph platform is isobutanol, a fuel for gasoline blending which has several positive attributes. Isobutanol is a clean burning and non-corrosive "drop-in" fuel which can be blended with gasoline and is compatible with the existing petroleum infrastructure. Currently, provisions allow for bio-butanol blending up to 12.5% with gasoline and ASTM specifications exist for sale as fuel. Every 1% of isobutanol blended with gasoline presents an $8 billion market opportunity. Combustion of isobutanol has cleaner emissions compared to gasoline by reducing hydrocarbon, carbon monoxide, nitrous oxide and sulfur dioxide production. Isobutanol also has advantages over first generation biofuels, such as ethanol, because of its higher energy density.

Intrexon Energy Partners was formed as a joint venture to commercialize the production of isobutanol from natural gas using a methanotroph. Gas fermentation processes were developed to evaluate growth of the bacteria and production of isobutanol from the microfermenation scale through the lab and pilot scales. The ability of our modified methanotroph to produce isobutanol has increased exponentially with time as the genetic toolbox was developed and early learnings in strain engineering were optimized into a design-build-test-learn cycle. A state of the art laboratory was custom built to bring research and development from a low throughput and manual process to a high throughput automated process. Increases in automation and throughput have positioned Intrexon as a leader in the genetic engineering of methanotrophs and accelerated the commercialization timelines of other chemical products.

To successfully bring this technology to commercial scale, consideration must be given not only to optimizing the metabolic pathway engineered into the methanotroph, but also to key issues in the fermentor design and downstream processing. Laboratory scale learnings were applied to select technology that was commercially feasible at demonstration scale. Since the substrate is a gas, the design of the fermentor must be optimized for effective mass transfer of the natural gas to the liquid phase for increased bioavailability. Downstream processing must take into consideration purification of the isobutanol and removal of fermentation byproducts and contaminants. Process selection must weigh the efficiency and selectivity of technology options, while at the same time, utilize as much mature technology as possible to minimize risk. The project team has performed numerous design reviews known as Front End Loading (FEL) to develop process criteria and scenarios

Beyond isobutanol, Intrexon's versatile platform strategy has accelerated the development of 2,3-butanediol, 1,4,butanediol and isobutyraldehyde platforms. Production of these intermediate chemicals has expanded the value of the natural gas conversion technology from the fuels market to consumer markets like tires and textiles. The commercial fermentation design optimization occurring in the pilot plant may also be applicable to other chemicals, shifting the focus of future products to downstream development.

495331 Comparison of Conventional and Newly Proposed Solvent Deasphalting Processes: Numerical Simulation Based on Energy and Cost Analysis

Abstract:

Oil consumption in non-OECD countries has been increasing, while the existing reserves of conventional oil keep decreasing. To replace conventional oil, many attempts have been made to use unconventional oil such as oil sand bitumen and extra heavy oil, which are difficult to extract from a reservoir. Unconventional oil requires upgrading processes because its asphaltene compounds containing a lot of sulfur and heavy metals are much higher than those of conventional oil. Among various upgrading processes, solvent deasphalting (SDA) process has the advantage of lower operating cost than other oil conversion processes and high quality deasphalted oil (DAO) can be obtained by selectively rejecting asphaltene in the SDA process. In the SDA process, about 3â€“5 times more solvent than feed is used to extract DAO and the solvent recycling is the most energy consuming part. In this study, we propose a novel process that consumes less energy than the conventional SDA process. In the proposed new process, the temperature of the first separation column is adjusted to efficiently operate a heat exchanger and carbon dioxide is used as a stripping agent to separate the solvent from DAO in the last separation column. To compare energy consumption and cost of the conventional and proposed SDA process, a commercial software was used for numerical simulation. The energy consumption was calculated by a graphical method using composite and grand composite curves. After calculating minimal energy consumption, operating and capital costs were obtained with changing the type of hot and cold utilities. The proposed new SDA process can produce the same quality and quantity of DAO as the conventional SDA process, but its energy consumption and total cost are lower than those of the conventional process.

502544 Modeling and Reaction Kinetics of Reactors for Polyesters Production

Abstract:

Scale-up of polymerization processes for the production of co-polyesters is complicated by the wide range of raw material units that can be used. For this reason, the knowledge of kinetics, mass transfer and thermodynamic characteristics of the reaction medium are important for the design of new polymers.

In this paper, an approach for the determination of reaction kinetics at the laboratory scale will be presented. The results will then be used in the reactor models for the scale-up of new polymer recipes to the commercial scale reactor.

A comparison of the modeling and experimental data at different scales will be finally presented.

490805 Modeling and Simulation of Nonwovens Manufacturing

Abstract:

It is of interest to increase the efficiency of fiber web materials used in products ranging from filters to thermal and acoustic insulators. Particular areas of optimization include throughput increase, final product quality and waste reduction. Both analytical modeling and Computational Fluids Dynamics (CDF) simulations are used as tools to optimize nonwovens manufacturing. Analytical modeling is based on a transport phenomena approach using the momentum, continuity and heat transfer equations. Multiphase CFD simulations using open source software and internally developed codes are used to simulate processing conditions and conduct optimization while accounting for turbulence and polymer rheology. Material properties and process conditions are henceforth tailored in the quest of optimum webs.

SOME STUDIES ON THE INHERENT RESILIENCE FOR A GAS SWEETENING UNIT

Sirshendu Guha, Engineers India Limited, New Delhi, India

Abstract

Mitchell and Mannan (2007) evaluated resilience of simple process systems like steam pipe, water pipe, water pump and heat exchanger etc., and considered physical failures of these systems. However, they did not provide any mathematical background and their approach was too simplistic. Hence, it could not be directly applied to complex process systems such as absorber, regenerator and distillation columns. Linh et al. (2012) reported resilience analysis considering physical failure of the process systems to improve the safety of process plants.

In contrast to physical failure of process systems, quantification of system resilience has been carried out considering performance failure of process systems under this work. The inherent resilience for process systems can be conceptualized from the perspectives of material resilience.

Correlations have been developed to assess resilience properties of constituent process systems pertaining to a Gas Sweetening Unit (GSU) as a case study. Fig.1 shows the schematic flow diagram of the Gas Sweetening Unit (GSU).

A steady state condition has been considered and system stress and system strain equations have been developed to quantify the system resilience.

This study has enabled developing a new method for identification of thermodynamically efficient operating capacity regime for optimization of energy usage in process systems. Optimum energy usage for a process system is achieved when it is operated at the most thermodynamically efficient capacity regime and this is evaluated by estimation of newly conceived Thermodynamic Coefficient of Performance (THCOP) data at various capacities.

This work provides general formulae for quantification of system resilience properties of one absorber column, one regenerator column, one shell and tube heat exchanger and associated one process as well as one utility pipeline system pertaining to a GSU (Fig.1).

It is found that the GSU (Fig.1) operating capacity range of 80%-100% is the most thermodynamically efficient operating capacity regime. The minimum operating capacity for the GSU is considered as 1% whereas the maximum operating capacity is 150%. The design capacity (i.e., 100%) of the GSU is 27814 kg/hr. The GSU is designed to sweeten sour natural gas containing 1.28 wt% of H_2S to sweet natural gas containing less than or equal to 1 ppmw H_2S.

It is found that absorber column and regenerator column systems under study possesses inherent resilience of around 5% and 15% respectively with regard to variation in upstream feed sour gas flow rate beyond 100% design flow rate, i.e., 27814 kg/h.

It is also found that the lean-rich exchanger system under study possesses inherent resilience of around 10% with regard to variation in upstream feed sour gas flow rate beyond 100% design flow rate.

It is noticed that the inlet feed sour gas pipeline system (process pipeline) indicates drastic fall in modulus of resilience value (*Ur*) and great increase in modulus of elasticity (*Esys*) value above feed gas mass flow rate of 100% indicating attainment of allowable pressure drop and limiting or yielding stress.

It is also found that the utility pipeline system (which carries superheated steam) shows a plateau at very high steam flow rate (57600 kg/hr-64800 kg/hr) indicating attainment of allowable pressure drop and limiting or yielding stress region for the system in case of flow variation.
Results also indicate that similar to a material, all the process systems under study (i.e., absorber, regenerator, lean-rich exchanger, process pipeline and utility pipeline) of Gas Sweetening Unit(GSU, Fig.1) demonstrate inverse relationship of modulus of resilience (*Ur*) with modulus of elasticity (*Esys*) in all applicable operating variable deviation regimes.
Computer simulation using a process simulator SIMULATION SCIENCES INC, Pro/II has been utilized for this study.
Finally, one example is given regarding design procedure in relation to incorporation of 50% over capacity factor or inherent resiliency in the absorber column by augmentation of number of column trays.

Introduction

Quantification of "resilience" in the domain of process systems is still elusive. Fiksel (2006) understood "system resilience as the capacity of a system to tolerate disturbances while retaining its structure and function" (Fiksel, 2006).Material resilience is the extent to which energy may be stored in a material by elastic deformation (Mitchell and Mannan, 2007). Quantification of resilience for a material can be easily done by evaluating the area under the stress-strain curve from zero stress value to the yield stress value (Mitchell and Mannan, 2007).

Likewise a material, resilience of a process system may also be defined as the quantity of energy, the system will be able to store before attaining the "point of instability" (Mitchell and Mannan, 2007).System instability may be understood as inability of the system to perform efficiently at its targeted performance levels.

Flexibility and resiliency of process systems were first described by Morari (1983). Guha and Das (2014) reported that in an insulated pipe segment system which carries superheated steam as process fluid, magnitudes of inherent system resiliencies are 927.8 kJ/m^3-s, 43 kJ/m^3-s and 31.5 kJ/m^3-s for variation of mass flow rate, inlet pressure and inlet temperature respectively.
In this work, the resilience magnitudes of process systems have been quantified by evaluation of the area under the system stress -system strain curve from the system stress value corresponding to a minimum operating condition to the system stress value corresponding to a maximum allowable operating condition. System stress corresponding to a maximum allowable operating condition will be limited by the design condition up to which the system will be able to maintain its targeted performance levels.

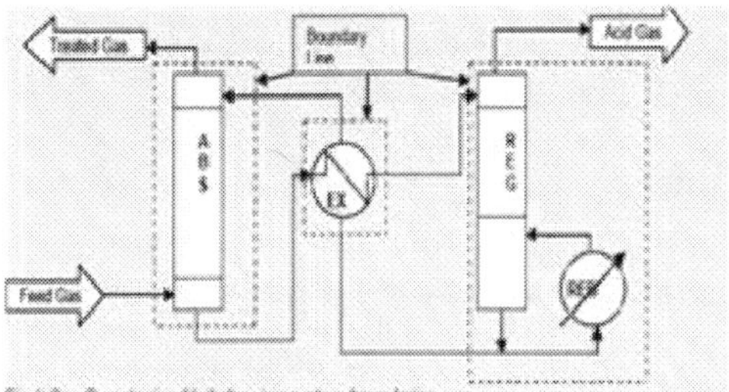

Fig 1 Gas Sweetening Unit showing system boundaries

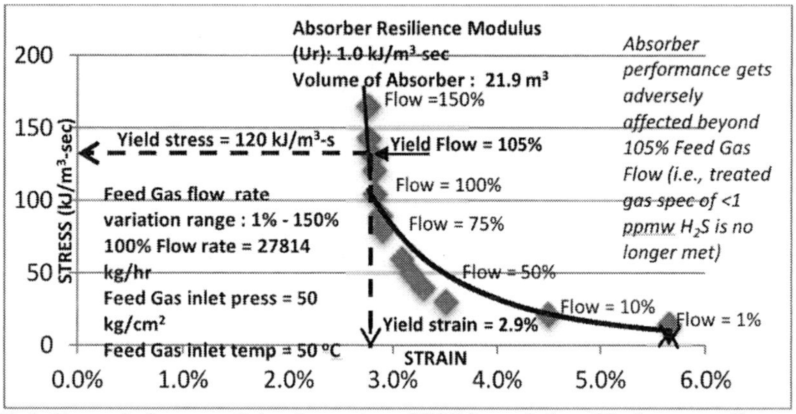

Fig.2: Stress-strain curve of absorber system of GSU for variation of feed gas mass flow rate

Research Objectives

In this work, following research objectives have been envisaged.

(a) Review of mathematical equations that can be used to assess resilience properties of a material.
(b) Development of mathematical equations for quantification of resilience properties for complex process systems pertaining to a Gas Sweetening Unit (GSU, Fig.1).
(c) Determination of thermodynamically efficient operating capacity regime for a GSU.

Methodology

(a) Review of mathematical equations that can be used to assess resilience properties of a material.

The amount of work which is required to deform unit volume of a material to its elastic limit was considered as the resilience of the material (Mitchell and Mannan, 2007). For a linearly elastic material, the modulus of resilience (U_R) can be represented as (Mitchell and Mannan, 2007).

$$U_R = \Omega^2_{yield} / 2E \qquad (1)$$

In (Eq.1), Ω_{yield} is yield stress of the material and E is modulus of elasticity of the material.

Development of (Eq.1) is based on the assumptions that elongation of the material has been carried out in an ideal reversible way maintaining equilibrium at all times and the cross-sectional area of the material does not alter during the process of elongation.

Since actual elongation process can never be carried out following an ideal reversible path and the cross-sectional area of the material changes during elongation process, one can construe that the area under the curve of *true material stress* and *true material strain* from zero stress to the yield stress; should be used for determination of the modulus of resilience (U_R) of the material.

If we consider a rod of length 'L' and 'A' cross section which is elongated, with 'X' representing the extent of elongation, we can state that for infinitesimal elongation 'dX', the infinitesimal work done 'dW' on the rod (-ve work as per convention) can be represented as,

$$dW = -\Omega.A.\ dX = -\Omega\ dV \qquad (2)$$

In the above equation, 'Ω' is material stress; 'V' is material volume.

From (Eq.2), one can define *true material stress* 'Ωt' in terms of irreversible or actual work $W(i)$ & change in material volume (ΔV) as ,

$$\Omega t = -dW/dV = dU/dV = -W(i)/\Delta V\ (dU = -dW \text{ when } dQm = 0) \qquad (3)$$

$$\text{Or } \Omega t = -\text{Actual Work/Volume Change} \qquad (3a)$$

In (Eq.3), 'Qm' is heat transferred into the material; 'U' is internal energy of the material.

Using 1st and 2nd laws of thermodynamics and experimental results (Raghavan, 1990) one can obtain following relation for lost work pertaining to a material during the process of elongation (Ta being environment temperature, S being entropy of the material, F being force applied on the material).

$$F = -Ta\ (dS/dX)_{\text{at constant } Ta} \qquad (3b)$$

$$\text{Or } -Ta\ dS = FdX = \Omega t\ .A.dX \qquad (3c)$$

$$\text{Or } d(X/L) = -TadS/\ \Omega t.A.L \qquad (3d)$$

$$\text{Or } d\varepsilon t = -TadS/\ \Omega t.V = -TadS\ (dV/V)/\ dU \qquad (4)$$

True material strain is 'εt' which can be defined in terms of lost work $W(L)$, fractional volume change (C) & actual work $W(i)$ as,

$$\varepsilon t = C\ (-TadS\ )/\ dU = C\ W(L)/-W(i) \qquad (5)$$

$$\text{Or } \varepsilon t = [\text{Vol.Change Fraction}]\ x\ [\text{Lost Work}]/[-\text{Actual Work}] \qquad (5a)$$

In (Eq.5), $C = dV/V$, which is a dimensionless quantity, and is not equal to dL/L as cross-sectional area changes during an elongation process. The entropy change is due to change in configuration of the material during the process of elongation. For non-linearly elastic materials, one can use values of true stress, computed from (Eq.3) and values of true strain εt, computed from (Eq.5) for graphical evaluation of material modulus of resilience 'U_R'. Using (Eq.3) and (Eq.5), "U_R" for a linearly elastic material which obeys the yield stress-strain relation:

$\Omega_{yield} = E\ \varepsilon_{yield}$, can be expressed as,

$$U_R = [-C\ Ta/2\ dS/dV]_{yield} = [\ Ta\ (S1-S2)/2V1]_{yield} = \Omega^2_{yield}/\ 2E \qquad (6)$$

In (Eq.6), S1, S2 are initial and final entropies of the material. V1 is initial material volume.

(b) Development of mathematical equations for quantification of resilience properties for complex process systems pertaining to a Gas Sweetening Unit (GSU, Fig.1)

System Stress: In line with (Eq.3) of a material, the stress equation of a system for unsteady flow condition can be derived by using the first law of thermodynamics as (Guha and Das, 2015),

$$S_{su} = 1/V_{sys}[d(MU)_{sys}/dt] = [\sum(mh)_i + \sum(mu^2/2)_i + \sum(mgz)_i + \sum q_j - \sum w_l]/V_{sys} \quad (7)$$

Similarly, for steady flow condition, the system stress equation can be written as (Guha and Das, 2015),

$$S_{ss} = [\sum(mh)_{in} + \sum(mu^2/2)_{in} + \sum(mgz)_{in} + (\sum q_j)_{in} - (\sum w_l)_{in}]/ V_{sys} \quad (8)$$

System Strain: Likewise a material (Eq.5), system strain can also be viewed as response of the system to application of a load in terms of change in its original entropy and is necessarily dimensionless. One can define system strain or response as ratio of "Waste work" to "Exergy$_{in}$" (input exergy) and the general equation for system strain or response can be represented as (Guha and Das, 2015),

$$S_{sn} = T_o\beta/\text{Exergy}_{in} = \text{Waste Work}/\text{Exergy}_{in} \quad (9)$$

where $T_o\beta = \sum[m(h-T_os]_i + \sum(1-T_o/Tj)q_j - \sum w_l$ = Waste Work

Based on the system stress and system strain definitions given above, correlations are developed for the quantification of system resilience modulus (Ur) for constituent complex process systems (absorber, lean-rich exchanger & regenerator) of a Gas Sweetening Unit (Fig.1, Fig.2):

System Stress Equations:

$$S_{s(abs)} = [\sum(mh)_{in}]/V_{(abs)} = 1/ V_{(abs)}[L_{sl}h_{sl} + V_{fg}h_{fg}] \quad (10)$$

$$S_{s(LRRxch)} = [(mh)_{in}]/V_{(LRExch)} = 1/ V_{(LRExch)}[L_{sr}h_{sr} + L_{sl}\, h_{sl}]_{at\ inlet} \quad (11)$$

$$S_{s(reg)} = [(mh)_{in} + \sum q_j]/V_{(reg)} = 1/ V_{(reg)}[L_{sr}h_{sr} + Q_{reb}] \quad (12)$$

System Strain Equations:

$$S_{sn(abs)} = [L_{sl}b_{sl} + V_{fg}b_{fg}] - [V_{tg}b_{tg} + L_{sr}b_{sr}] / L_{sl}[(h_{sl} - h^0{}_{sl}) - T_0(s_{sl} - s^0{}_{sl})] + V_{fg}[(h_{fg} - h^0{}_{fg}) - T_0(s_{fg} - s^0{}_{fg})] \quad (13)$$

$$S_{sn(LRExch)} = T_0\, [L_{sr}\, (s_{srout}\text{-}s_{srin}) + L_{sl}(s_{slout}\text{-}s_{slin})] / L_{sr}[(h_{sr} - h^0{}_{sr}) - T_0(s_{sr} - s^0{}_{sr})] + L_{sl}[(h_{sl} - h^0{}_{sl}) - T_0(s_{sl} - s^0{}_{sl})] \quad (14)$$

$$S_{sn(reg)} = [L_{sr}b_{sr}] - [\, V_{ag}b_{ag} + L_{sl}b_{sl}] + (1\text{-}T_0/T_{reb})Q_{reb} / \; L_{sr}[(h_{sr} - h^0{}_{sr}) - T_0(s_{sr} - s^0{}_{sr})] + (1\text{-}T_0/T_{reb})Q_{reb} \quad (15)$$

System resilience modulus (Ur) can be quantified by evaluation of the area under the system stress - system strain curve from the system stress value corresponding to a minimum operating condition to the system stress value corresponding to a maximum allowable operating condition. Soft computing simulation technique, using a process simulator needs to be utilized for generation of system stress-strain data for varying operating conditions.

Process Simulation

Soft computing process simulation technique has been used utilizing a high accuracy SIMULATION SCIENCES INC process simulator, Pro/II for this study. The process simulator has been used for steady state simulation of the Gas Sweetening Unit (GSU) that has been designed to sweeten natural gas which contains 1.28 wt% H_2S. Gas sweetening is accomplished by usage of Di-Ethanol amine as solvent. The unit is designed for a treated gas H_2S specification of < 1 ppmw. The absorber and regenerator columns of the GSU have been simulated considering equilibrium stages. The chosen thermodynamic model which has been considered for the entire simulation input file is "special amine package".
The physical and thermodynamic property data of all process streams (external and internal of absorber column, regenerator column and external of lean-rich exchanger) generated in the process simulator output files are utilized in the quantitative correlations which are developed and used in order to estimate variation of stress and strain with varying operating conditions for the absorber column, regenerator column and lean rich exchanger systems.

Results & Analysis

Fig.2 indicates that the absorber column system under study possesses inherent resilience of around 5% with regard to variation in upstream feed sour gas flow rate beyond 100% design flow rate, i.e., 27814 kg/h. Hence, this absorber column may be designed for 95% feed gas capacity even if the column is actually operated for 100% feed gas capacity. Conversely, if the column is designed for 100% feed gas capacity, it may be actually operated at 105% feed gas capacity. The absorber column system is a tray column of 16.5m height (TL-TL) and 1.3m (1300 mm) internal diameter corresponding to 100% feed gas capacity.

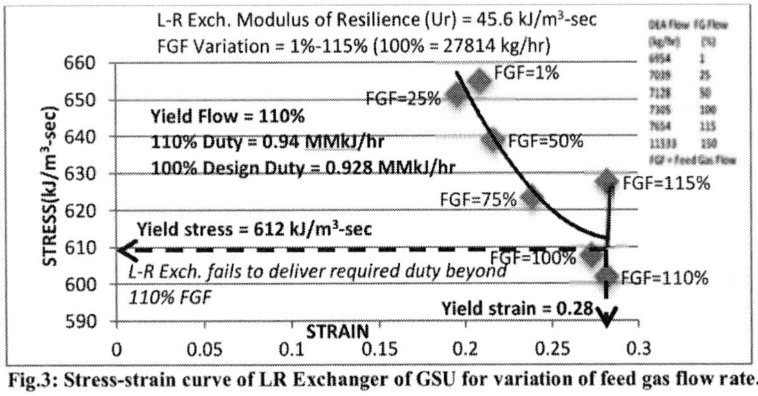

Fig.3: Stress-strain curve of LR Exchanger of GSU for variation of feed gas flow rate.

Fig.3 shows that the lean-rich exchanger system under study possesses inherent resilience of around 10% with regard to variation in upstream feed sour gas flow rate beyond 100% design flow rate. Similar to the absorber column, the lean-rich exchanger may be designed for 90% feed gas capacity although it is actually operated for 100% feed gas capacity. In other words, the lean-rich exchanger may be operated for 110% feed gas capacity even if it is designed for 100% feed gas capacity. In case of 100% feed gas capacity, the designed lean-rich exchanger is a TEMA AES type (1- shell pass and 2-tube passes) heat exchanger with design duty (for 100% feed gas capacity) of 0.928×10^6 kJ/hr corresponding to design U*A (overall heat transfer coefficient being U and heat transfer surface area being A) of 4830 kJ/hr-°C with MTD (mean temperature difference) of 192.2 °C.

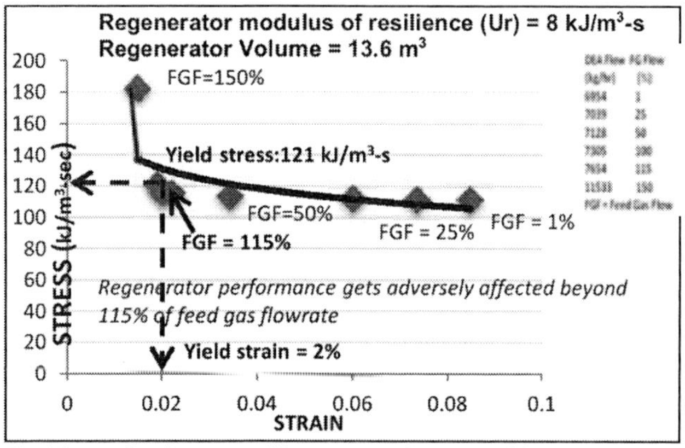

Fig.4 Stress-Strain curve of Regenerator System of GSU for variation of feed gas flow rate

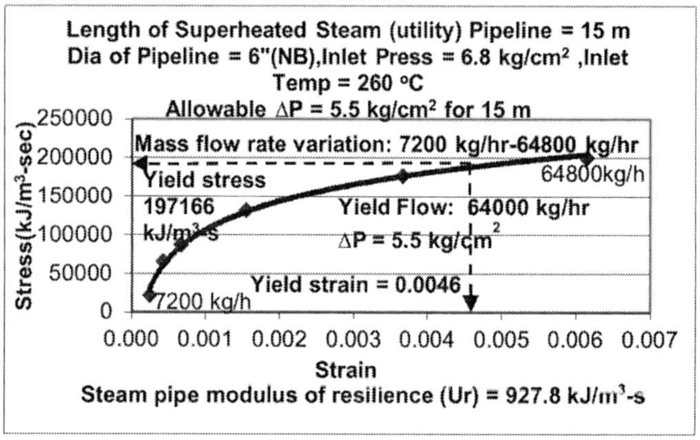

Fig.5 Stress-strain curve of superheated steam pipeline system for mass flow rate variation

Fig.4 indicates that the regenerator column system under study possesses inherent resilience of around 15% with regard to variation in upstream feed sour gas flow rate beyond 100% design flow rate, i.e., 27814 kg/h. The regenerator column may be designed for 85% feed gas capacity in spite of the same being actually operated for 100% feed gas capacity. Alternatively, the regenerator column may be operated for 115% feed gas capacity although it is designed for 100% feed gas capacity. The regenerator column system is a tray column of 21.5m height (TL-TL) and 0.9m (900 mm) internal diameter corresponding to 100% feed gas capacity.
It may be noted that since the limiting inherent resilience is 5% pertaining to the absorber column, overall, for the entire GSU, minimum inherent resilience or flexibility of 5% can be allowed even if the unit is designed for 100% design capacity (i.e., 27814 kg/hr).

Fig.5 shows one utility pipeline system (which carries superheated steam) stress versus strain for steam mass flow rate variation of 7200 kg/hr to 64800 kg/hr. The curve shows a plateau at very high steam flow rate region of 57600 kg/hr to 64800 kg/hr indicating attainment of practically allowable maximum pressure drop of 5.5 kg/cm^2 pertaining to 15 m pipeline length at yield or limiting stress of 197166 kJ/m^{3}s corresponding to yield or limiting flow of 64000 kg/hr for the utility pipeline system in case of flow variation. The physical significance of attainment of the yield or limiting flow of 64000 kg/hr is that the 6" (NB) pipeline system will not be able to carry superheated steam of more than 64000 kg/hr as long as inlet pressure of this steam is fixed at 6.8 kg/cm^2.

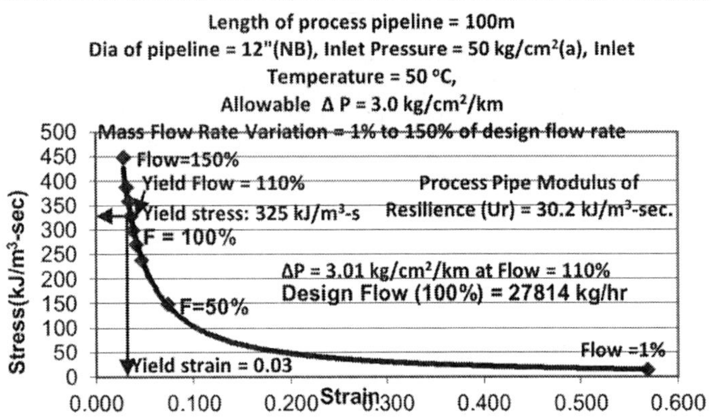

Ur (kJ/m³-s)	*Esys (kJ/m³-s)*	*Flow range (%)*
19.41	272	1-50
7.4	4175	50-100
2.27	**13382**	**100-130**

Fig.6 Stress versus strain curve for inlet sour feed gas pipeline system for mass flow rate variation

Fig.6 shows the inlet sour feed gas pipeline system (process pipeline) stress versus strain for deviation in mass flow rate of feed sour gas from 1% of normal throughput to 150% of the normal throughput (100% normal throughput = 27814 kg/hr). The curve indicates drastic fall in modulus of resilience value (*Ur*) and great increase in modulus of elasticity (*Esys*) value at mass flow rate variation range of 100% -130% of feed gas indicating attainment of allowable pressure drop of 3.0 kg/cm^2/km at yield or limiting stress of ~ 325 kJ/m^{3}s. Allowable pressure drop of 3.0 kg/cm^2/km is a defined limiting parameter based on which the pipeline size of 12" (NB) has been computed pertaining to 100% mass flow rate (27814 kg/hr) of feed gas. The curve also demonstrates inverse relationship of modulus of resilience (*Ur*) with modulus of elasticity (*Esys*) for the pipeline system of sour feed gas.

(c) Determination of thermodynamically efficient operating capacity regime for a GSU.

Optimum energy usage for a process system is achieved when it is operated at the best thermodynamically efficient capacity regime which also ensures operation in minimum waste work capacity range (Guha, 2015). Best thermodynamically efficient capacity regime can be evaluated by estimation of Thermodynamic Coefficient of Performance (THCOP) data at various operating capacities (Guha, 2015).

Thermodynamic Coefficient of Performance (THCOP) is defined as (Guha, 2015),

$$THCOP = [1 + \beta/\sum (m_i s_i)_{in}] \qquad (16)$$

Where, β = rate of entropy creation in a system and $(m_i s_i)_{in}$ = rate of entropy inflow with ith stream.

If THCOP = 1.0, the system is best thermodynamically efficient. That is rate of irreversible entropy creation (β) = 0. Waste work for a separation process can be evaluated as (Guha, 2015),

$$(T_0\beta) = \text{Waste work} = \text{Total work} - \text{Minimum work} \qquad (17)$$

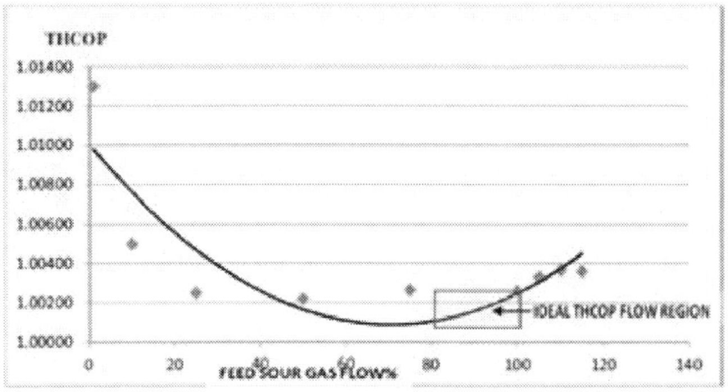

Fig.7 THCOP variation with Feed Sour Gas Flow%

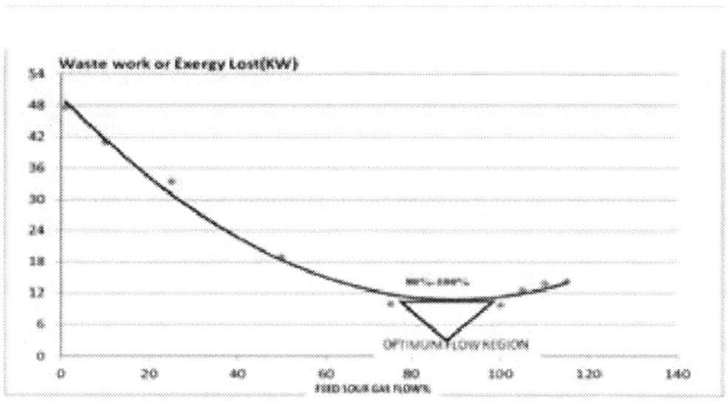

Fig.8 Waste work variation with Feed Sour Gas Flow%

In order to evaluate best thermodynamically efficient operating capacity regime of a GSU (Fig.1), THCOP (Fig.7) as well as Waste work (Fig.8) variation against variation in Feed sour gas flow rate pertaining to the regenerator column is to be determined as in a GSU (Fig.1), energy usage takes place in the regenerator column system only (steam is used in regenerator re-boiler).

Fig.7 and Fig.8 clearly indicate that the GSU operating capacity range of 80%-100% is the best thermodynamically efficient operating capacity regime. Moreover, since optimum or ideal flow region has been evaluated as 80%-100% (Fig.7, Fig.8) for the regenerator column and in turn for the entire GSU, the GSU turndown capacity should not be less than 80% of its design capacity. In other words, operation of the GSU below 80% of its design capacity will incur higher percentage of wastage of valuable work or energy in spite of spending lesser total work or energy at these capacities.

Variation of Modulus of Resilience (Ur) with Modulus of Elasticity (Esys)														
Absorber			Regenerator			Lean-Rich Exch.			Steam Pipeline			Sour feed gas Pipeline		
Ur	Esys	Flow range*	Ur	Esys	Flow range*	Ur	Esys	Flow range*	Ur	Esys x 10^6	Flow range#	Ur	Esys	Flow range*
0.739	1738	1-50	2.6	44.6	1-25	4.8	2137	1-50	28.49	154.8	7200-28800	19.41	272	1-50
0.184	22852	50-100	2.9	42.5	25-50	35.5	548	50-100	430	29.9	28800-57600	7.4	4175	50-100
0.077	73844	100-150	1.5	195	50-100	5.3	112391	100-115	469	9.3	57600-64800	2.27	13382	100-130
			0.5	1760	100-115									
			0.5	14520	115-150									
*Ur in kJ/m³-s, Esys in kJ/m³-s, *Flow range in %, #Flow range in kg/hr*														

Application to Process Design

Design procedure in relation to incorporation of 50% over capacity factor or inherent resiliency in the absorber column by augmentation of number of column trays can be found in literature (Guha and Das, 2015).

Conclusions

Inherent resilience for process systems have been conceptualized from the perspectives of material resilience for development of mathematical equations which are used to assess resilience properties of constituent process systems pertaining to a GSU (Fig.1).

The study has enabled developing a new method for identification of thermodynamically efficient operating capacity regime for optimization of energy usage in a GSU. GSU operating capacity range of

80% to 100% has been determined as the optimum operating capacity regime with regard to energy usage for its operation. The frame work of application of resilience analysis into process equipment design with regard to identification of thermodynamically efficient operating capacity regime for the purpose of energy usage optimization is applicable to any process system as well as process unit. Furthermore, usage of this concept may provide roadmaps for minimizing CAPEX (capital expenditure) at the cost of little enhancement in OPEX (operating expenditure) for any process plant in CPI.

Notation

A = surface area of L-R exchanger,
b_{sl} = availability function of lean solvent,
b_{tg} = availability function of treated gas,
b_{ag} = availability function of acid gas,
g_i = stream i acceleration due to gravity,
h_{sr} = enthalpy of rich solvent,
h^o_{fg} = enthalpy of feed gas at datum cond.,
h_i = stream i enthalpy,
h^o_{sl} =enthalpy of lean solvent at datum cond.
L_{sr} = rich solvent flow rate,
M_{sys} = system mass,
m_i = stream i mass flow rate,
q_{jin} = heat input rate,
S_{su} = system stress (unsteady state),
S_{sn} = system strain (steady state),
$S_{sn(reg)}$ = regenerator strain,
s_i = stream i entropy,
$S_{s(LR\ Exch)}$ = Lean-Rich Exch. stress,
s_{sl} = entropy of lean solvent,
s_{fg} = entropy of feed gas,
s_{srout} = entropy of out flowing rich solvent,
s_{slout} = entropy of out flowing lean solvent,
s_{sr} = entropy of rich solvent,
t = time,
T_o = environment/datum temperature,
U = over-all heat transfer co-eff. of L-R Ex.
u_{in} = input stream velocity,
Ur = system modulus of resilience,
V_{fg} =feed gas flow rate,
V_{reg} = regenerator volume,
V_{sys} = system volume,
V_{ag} = acid gas flow rate,
wl = work per unit time for part of system,
z_{in} = input stream elevation,
$b = (h-T_os)$ = availability function.
b_{fg} = availability function of feed gas.
b_{sr} = availability function of rich solvent.
$Exergy_{in}$ = exergy input.
g_{in} = input stream acceleration due to gravity.
h_{fg} = feed gas enthalpy.
h_{in} = enthalpy of input stream.
h_{sl} = enthalpy of lean solvent.
h^o_{sr} = enthalpy of rich solvent at datum cond.
L_{sl} = lean solvent flow rate.
m_{in} = input stream mass flow rate.
Q_{reb} = reboiler duty.
q_j = heat exch. rate between system & external reservoir.
S_{ss} = system stress (steady state).
$S_{s(reg)}$ = regenerator stress.
$S_{sn(abs)}$ = absorber strain.
$S_{s(abs)}$ = absorber stress.
$S_{sn(LR\ Exch)}$ = Lean-Rich Exch. strain.
s^o_{sl} = entropy of lean solvent at datum conditions.
s^o_{fg} = entropy of feed gas at datum conditions.
s_{srin} = entropy of in flowing rich solvent.
s_{slin} = entropy of in flowing lean solvent.
s^o_{sr} = entropy of rich solvent at datum conditions.
T_j = environment/external reservoir temperature.
T_{reb} = reboiler temperature.
u_i = stream i velocity.
U_{sys} = system internal energy.
$Esys$ = system modulus of elasticity.
$V_{(LR\ Exch)}$ = Lean-Rich Exch. volume.
$V_{(abs)}$ = absorber volume.
V_{tg} = treated gas flow rate.
wl_{in} = work input.
z_i = stream i elevation.
β = rate of entropy creation.

References

1. Fiksel, J., "Sustainability and Resilience: Towards a systems approach", ***Sustainability: Science, Practice & Policy, 2, 14-21,(2006).***
2. Guha, S., and Das, S.K., "Resilience Analysis of a Pipe Segment System Carrying Superheated Steam", ***Advanced Materials Research, 917, pp. 232-243, (2014).***
3. Guha, S. and Das, S.K., "Thermodynamic Availability Analysis: Usage in Quantification of Resilience Modulus, Modulus of Elasticity and Yield Stress for an Absorber Column System", ***IJCSIT, Vol. 6(4), (2015),3630-3649.***
4. Guha, S., "Simulation Studies for Analysis of Inherent Resilience Properties of an Amine Regeneration Column Used in a Gas Sweetening Unit", ***Procedia Computer Science, 62 (2015), 593-602.***
5. Linh T.T. Dinh, Hans Pasman, Xiaodan Gao, M. Sam Mannan (2012), "Resilience engineering of industrial processes: Principles and contributing factors", ***Journal of Loss Prevention in the Process Industries, 25,233-241.***
6. Mitchell S. M., M. Sam Mannan., Ph.D. Thesis, "Resilient Engineered Systems: The Development of an Inherent System Property", ***Texas A&M University, Texas (2007).***
7. Morari, M., "Flexibility and resiliency of process systems", ***Comp. Chem. Engg., 7, 423-437,*** **(1983).**
8. V. Raghavan (1990), "Materials Science and Engineering", ***Prentice Hall, 3rd Edition.***

491609 Tubular Multi-Tube Cooled Reactor As a Fractional Heat Exchanger - Modeling Reactive Heat Transfer

Abstract:

Many industrially important vapor phase reactions are equilibrium controlled and highly exothermal in nature which leads to high reactor exit temperature. Very often product gas requires cooling in a tubular heat exchanger before it is sent to the next unit operation. Residual reaction could occur under reduced temperature within the heat exchanger and end product profiles and final gas conditions are determined by the relative rates of heat transfer and reaction in such a unit. If the gas final cooled product compositions and conditions are important to the next unit operation, proper prediction, monitoring and control of the unit through rigorous process modeling is critical.

If the inlet gas to heat exchanger temperature is extremely high, then heat transfer by radiation should be considered since the effect of this heat transfer mode cannot be neglected when inlet gas temperature is very high. Exchanger outlet gas temperature can be controlled by properly design the residence time of gas inside the exchanger. Any reactions after the cooling require certain gas compositions and temperature which should be properly handled together with the gas cooler. Optimization of the combined unit operations will result in potentially reduced capital investment and operational costs, along with obtaining more desired final product compositions and conditions.

If the purpose of the cooling is purification of the gas stream, reactions occurring in either vapor or condensed liquid phase should be evaluated carefully to identify the extent of reaction and its impact on the separation performance. When a phase change exists, both phase and chemical equilibrium should be considered and mass transfer across the interface should be included along with heat transfer. Optionally chemical reaction kinetics and fluid flow of the condensed liquid should be considered.

This talk will present rigorous modeling of reactive heat transfer which may be useful to process intensifications.

494706 Thermal Stability of Ethylene Polymerization in Fluidized-Bed Reactor

Abstract:

Ethylene polymerization in fluidized bed reactor is a simple process and is easier for production separation. Thus fluidized bed reactor is widely used in ethylene and propene polymerization. In order to improve the production rate of polyethylene in fluidized bed reactor, condensed mode operation and super-condensed mode operation are developed. With inert condensed compound such as i-pentane or n-hexane or comonomer such as n-hexene, the recycle stream can become a gas-liquid mixture under the reasonable temperature, thus more heat of reaction can be removed by the recycle stream and the production rate will be dramatically increased.

Ethylene polymerization is a rapid and highly exothermic process. As the melting temperature of polyethylene near the reaction temperature, thermal instability of fluidized bed may cause the oscillation of reactor temperature, and even induce polyethylene melting and plant stop. Therefore, thermal stability and dynamical behavior of fluidized bed not only concern design and scale up of fluidized bed reactor, but also concerns the real time monitoring, handling of abnormal conditions and safety production.

Several methods have been employed to analyze the stability and dynamic behaviors, such as mathematics deduction method, nonlinear analysis method, process simulation method, etc. The mathematics deduction method is based on strict mathematical derivation, however, it cannot be used to solve relatively high dimensional systems. Nonlinear analysis method is used to solve relatively high dimensional systems. Meanwhile, the process simulation method is applied to solve large complex chemical process but this method still exist some disadvantages in stability analysis. In this work, a simple model of fluidized bed reactor is proposed to get the thermal stability criterion. Then effects of structural parameters of fluidized bed and operation parameters on thermal stability of ethylene polymerization process are investigated based on a detailed mathematical model of fluidized bed.

In the simple model of fluidized bed reactor, concentration of reactants and reactor pressure are assumed to be constant. Therefore, the simple model is a first order ordinary differential equation group with respect to catalyst concentration and reactor temperature. With mathematics deduction, the coefficient matrix of and thermal stability criterion are obtained. The calculation results show that space-time yield of polyethylene on the Hopf bifurcation points is proportion to catalyst activation energy, reactor height and square of reaction, and is in inverse proportion to superficial velocity and density of recycle gas.

Based on the two phase model, a multi zone model is proposed in order to describe ethylene polymerization in fluidized bed in condensed mode and super-condensed mode. In the multi zone model, fluidized bed is composed of emulsion phase and bubble phase, and the emulsion phase is divided into gas-liquid-solid zone and gas- solid zone. The evaporation of liquid occurs in the gas-liquid-solid zone with a certain thickness liquid film on the surface of solid particles. Furthermore, the liquid evaporation rate obeys the equilibrium evaporation principles. Then the structural parameters and operation parameters of fluidized bed reactor are chosen to trace solution curves and bifurcation points referred to the catalyst feed rate.

The calculation results of multi zone model show that there are Hopf bifurcation points and limit points in the solution curves, which divide the solution curve into three parts-stable part with low temperature, unstable part and stable part with high temperature. Moreover, the steady state multiplicity of fluidized bed reactor occurs and the temperature range of stable part with low temperature shrinks with increase of activation energy of catalyst and condensed compounds in the recycle stream. In addition, thermal stability criterion based on a simple model fit well with the prediction of multi zone model.

492382 From Tool to Discipline- Developing a Model-Centric Process Engineering Culture

Abstract:

Process simulators and first-principles models are widely used in the commodity chemicals industry to drive process engineering and design activities. The applicability of these tools in the specialty chemicals and polymer industry is challenging due to inadequate fundamental thermodynamic and kinetic data for chemical systems of interest, complexity of chemistries practiced, utilization of batch and semi-batch operations and "non-traditional" unit operations and equipment. A successful application of a model-based approach can result in more robust scale-up efforts for new process development and provide fundamental insight to guide process optimization efforts for existing plant assets.

A model-centric approach aims to put first-principles models at the center of all process engineering activity. The layered approach involves integrating different sets of expertise such as applied thermodynamics and phase equilibria, experimental design, reaction engineering and computational chemistry with the goal of developing a core template for process development efforts and asset templates for plant support activities. The objective is to develop a fundamental understanding of existing assets and technologies practiced and to "de-risk" new process development. This presentation will touch upon the experience of developing such a program at SABIC Specialties.

492264 Scale-up Modelling and Analysis of a Pharmaceutical Crystallization Process

Abstract:

One of the great challenges in the design of industrial crystallization processes is to predict the influence of crystallizer scale on the process behavior and product quality attributes. Industrial crystallizers demonstrate far from well-mixed behavior. Local variations of critical process variables such as supersaturation, particle concentration, temperature profiles, etc. exist within industrial crystallizers. Consequently, crystallization process models that are based on the assumption of well-mixed systems are not fit for use in supporting optimal scale-up and design of industrial crystallizers. A more detailed insight into the mixing behavior and its consequences on local crystallization phenomena â€“such as a distributed parameter model- is needed in order to achieve reliable design, including an appropriate process scale-up.

This work presents an approach to develop a predictive model of a pharmaceutical crystallization process scale-up based on compartmental modelling. While full blown CFD models could be used to better describe and predict mixing characteristics in reactors as opposed to compartmental modeling, they are not practical for industrial applications due to heavy computational burden. Compartmental modelling is such that a trade-off approach is used to overcome the limitations of well-mixed models, by considering localized mixing, heat transfer and fluid hydrodynamics separately from crystallization kinetics within a crystallizer. While compartmental modelling is a well-known and successfully applied technique in other areas of process engineering, such as characterizing mass transfer for oxygen in industrial fermentation, there is a lack of comprehensive studies on the application of compartmental modeling approach to the scale-up of industrial pharmaceutical crystallization.

In order to apply compartmental modelling to pharmaceutical batch cooling crystallization, the crystallizer is divided into finite number of compartment volumes in a way that every compartmental zone possesses minimized or negligible gradients in profiles, such as crystal distribution, super-saturation and energy dissipation within their individual volumes. The same set of model equations and model parameters are defined in order to solve the balance equations for the conservation of population, mass and energy for every compartment. However, spatial variations in the crystallizer are modelled by different rates of nucleation, growth, dissolution and attrition in the individual compartments. The compartment modelling is implemented in MATLAB/Simulink.

The process behavior of the pharmaceutical batch cooling crystallization is analyzed with respect to crystallizer scale, process variables and operation conditions. The influence of the non-uniformly distribution of the related process variables is shown by comparing the multi-compartment model with a single-compartment well-mixed model.

Acknowledgment

We would like to thank the Danish Council for Independent Research (DFF) for financing the project with grant ID: DFF-6111600077B.

499257 Removal of Impurity from API and Reduction of Agglomeration during Drying

Abstract:

In this work we describe removal of an impurity from an API which was not detected in HPLC and can be observed only in NMR in specific solvents. A kinetic model was developed for conversion of this impurity to API using online NMR. Based on the prediction from the model the impurity was converted into API and isolated after crystallization. There were challenges in milling of the API due to large chunks (about 1 -2 cm) which were formed due to agglomeration during filtration and drying. This issue was resolved by using filter dryer, application controlled vacuum and application of agitation in the filter dryer based on LOD of the API during drying.

491244 Viscoelastic Properties and Dissolution Kinetics of Amorphous Drug Dispersion Films, Utilizing the Quartz Crystal Microbalance

Abstract:

Amorphous drug dispersions are becoming more and more utilized within the pharmaceutical industry, along with their academic study[1-2]. Their primary use is in facilitating the intake of poorly water soluble drugs by patients[3-4]. Plenty of research has been performed on understanding their physiochemical characteristics and stability in a wide range of conditions[2, 5]. Of the research published, little has gone into real time studies of the physiochemical properties of such films as they degrade. Our team, sought to utilize the Quartz Crystal Microbalance (QCM) to visualize how these drug blended films had altered physiochemical properties from unloaded ones in a variety of different environments[6].

Carbamazepine and Copovidone were the compounds used to form amorphous drug films that were spun cast onto the QCM sensors. Solution blends in Methanol of 17% Copovidone weight by volume and Carbamazepine drug loadings of 35 to 5 mg/mL were used and spun cast. The lack of crystallinity was proven by a combination of several instrumental techniques including optical microscopy, Powder X-ray Diffraction (PXRD), and Differential Scanning Calorimetry (DSC), and the presence of Carbamazepine was shown via Fourier transform infrared spectroscopy (FTIR).

Quantitative data, including the thickness, density, and viscoelasticity, on the structural properties of the film was obtained by utilizing both the Sauerbrey and Viscoelastic models for the films. The films were found to be better modelled as soft rather than rigid, and the Sauerbrey model was not as effective. The data obtained from the Viscoelastic model was corroborated with ellipsometry to obtain a thickness. Within this body of work, the films at varying temperatures and on different surfaces to analyse their effects on the physiochemical properties of the films.

Furthermore, the dissolution kinetics of these films when submerged in water and ethanol were investigated in real time. In water, depending on the drug loading, this led to the nucleation and crystallization of the Carbamazepine in the excipient film, where the contact of the particles on the sensors are seen. The particle sizing was performed using Dynamic Light Scattering (DLS) to obtain size measurements of the Carbamazepine powder that formed.

All these works combined, showcase the strength of the QCM as a surface specific tool and its potential within the pharmaceutical formulations industry in helping to optimize the combinations of excipients and active pharmaceutical ingredients. Additionally, it offers strong insight into how the properties of the amorphous drug films change with temperature. Future studies could utilize alternative sensors and cells to further expand our knowledge and understanding of these amorphous drug films.

References:

1. Baghel, S.; Cathcart, H.; O'Reilly, N. J. Polymeric Amorphous Solid Dispersions: A Review of Amorphization, Crystallization, Stabilization, Solid-State Characterization, and Aqueous Solubilization of Biopharmaceutical Classification System Class II Drugs. *Journal of Pharmaceutical Sciences* **2016,** *105*, 2527-2544.

2. Douroumis, D.; Bouropoulos, N.; Fahr, A. Physicochemical characterization of solid dispersions of three antiepileptic drugs prepared by solvent evaporation method. *Journal of Pharmacy and Pharmacology* **2007,** *59*, 645-653.

3. Alonzo, D. E.; Zhang, G. G. Z.; Zhou, D.; Gao, Y.; Taylor, L. S. Understanding the Behavior of Amorphous Pharmaceutical Systems during Dissolution. *Pharmaceutical Research* **2010,** *27*, 608-618.

4. Ehmann, H. M. A.; Winter, S.; Griesser, T.; Keimel, R.; Schrank, S.; Zimmer, A.; Werzer, O. Dissolution Testing of Hardly Soluble Materials by Surface Sensitive Techniques: Clotrimazole from an Insoluble Matrix. *Pharmaceutical Research* **2014,** *31*, 2708-2715.

5. Weuts, I.; Van Dycke, F.; Voorspoels, J.; De Cort, S.; Stokbroekx, S.; Leemans, R.; Brewster, M. E.; Xu, D.; Segmuller, B.; Turner, Y. T. A.; Roberts, C. J.; Davies, M. C.; Qi, S.; Craig, D. Q. M.; Reading, M. Physicochemical Properties of the Amorphous Drug, Cast Films, and Spray Dried Powders to Predict Formulation Probability of Success for Solid Dispersions: Etravirine. *Journal of Pharmaceutical Sciences* **2011,** *100*, 260-274.

6. Reviakine, I.; Johannsmann, D.; Richter, R. P. Hearing What You Cannot See and Visualizing What You Hear: Interpreting Quartz Crystal Microbalance Data from Solvated Interfaces. *Analytical Chemistry* **2011,** *83*, 8838-8848.

496430 Improving the Dissolution of Ibuprofen By Co-Milling of Ibuprofen and MCC Mixture in a Ball Mill

Abstract:

In this work, we investigated the improvement of dissolution of ibuprofen by co-milling the ibuprofen with microcrystalline cellulose, MCC (hydrophilic excipient) in dry mode. Ibuprofen and MCC were taken in 1:1 weight ratio and co-milled for different times in a planetary ball mill. Co-milled, v-blended and as received samples were characterized by different techniques. Particle size distribution measured using laser diffraction method exhibited a shift to lower particle size. To understand the morphology, shape and surface changes, optical microscopy and Field Emission Scanning Electron Microscopy (FE-SEM) were performed. While the crystallinity and crystal level changes (such as defects, strain energy and crystallite size etc.) were measured using Powder X-Ray Diffraction (P-XRD), the presence of different functional groups on the surface and the interactions between drug and excipient; and hydrogen bond formation was evaluated using Fourier Transform Infrared Spectroscopy (FTIR). The changes in the crystallinity due to co-milling were further confirmed using Differential Scanning Calorimetry (DSC) where the shift of melting endotherm to lower temperature was observed. Acidic and basic nature of the surface was evaluated by Inverse Gas Chromatography (IGC) measurements by acidic and basic parameters. The wettability measurements performed by the sessile drop method shows lower contact angle for co-milled samples demonstrating improvement in wettability of the ibuprofen. This is also confirmed by the dissolution test in USP Paddle II apparatus. The study shows that co-milling imparts surface hydrophilicity apart from size reduction by generating defects, dislocations and decreased crystallite size by the impact of balls on the surface. Increased interactions between carboxylic acid group and hydroxyl group result in hydrogen bond formation thereby improving the wettability and dissolution. Increase in polar surface energies and increase in the ratio of basic to acidic constant indicates hydrophilic nature of the surface which contributes to enhancing the dissolution of ibuprofen. Thus, co-milling can be an alternate solventless method to improve dissolution of BCS Class II drugs.

500110 Innovative Design and Systematic Process Intensification Using Building Blocks

Abstract:

Process intensification addresses the development of techniques that leads to substantially smaller, cleaner, safer, and more energy-efficient technologies. It also enables the innovative design of novel multi-functional units (e.g., Eastman Chemical methyl acetate process that performs extractive distillation, reactive distillation, reaction and distillation within a single column [1,2]). Intensification methodologies have been developed in the past, several of which are described in [3-8]. In this work, we address the following design and intensification challenges for both steady-state and dynamic processes: (i) how to systematically select and combine multiple phenomena to create new opportunities and obtain "out-of-the-box" design solutions, (ii) how to systematically identify intensified equipment and, at the same time, generate process flowsheets containing these equipment, (iii) how to select optimal configurations of intensified equipment and process flowsheets, and (iv) how to incorporate process intensification into process synthesis and optimization without *a priori* postulation of intensification alternatives. To address these questions within a single framework, we depart from the classical unit operation-based representation of process units, flowsheets and superstructures, and propose a new representation using finite building blocks [9]. In this representation, a physiochemical phenomenon, a processing task or a function can be represented using a single block or using two neighboring blocks and their boundaries. Other representations also exist in literature, including the modular [10,11] and phenomena building blocks (PBBs) [6,7]. Here, we propose a novel and general process superstructure which is an ensemble of only building blocks in two dimensions. Each block is denoted by its coordinate within the superstructure. A block can have entering and outgoing streams in both horizontal and vertical directions. An assembly of blocks of the same type obtains a classical unit, while an assembly of blocks with different types results in an intensified unit. The block superstructure encompasses numerous process equipment and flowsheet alternatives, and can be used to design, synthesize and intensify a chemical process without *a priori* postulation of these alternatives. The overall problem is formulated using a single mixed-integer nonlinear optimization (MINLP) model.

For dynamic intensification of multi-functional and periodic/cyclic processes, we propose a similar block superstructure, where each block represents a specific material performing a specific task for a specific time step. The superstructure is specifically suitable for the design of integrated, multi-functional and modular adsorption-reaction systems embedded within a single column while exploiting the dynamic interactions and trade-offs between different phenomena. Again, from the same superstructure, one can systematically generate variants of adsorption-reaction systems, including single- or multi-layered packed adsorption columns, reactors, pressure/temperature swing adsorption (PSA/TSA) processes, pressure and temperature swing adsorption (PTSA) processes, sorption-enhanced reaction processes (SERP), simulated moving bed chromatography (SMB) processes, and reactive SMB processes. Using this method, various intensified and modular processes can be designed for the utilization of both conventional and unconventional resources including stranded natural gas, contaminated shale gas, fuel gas or landfill gas to produce sustainable fuels and chemicals [12].

References:

[1] Agreda, V.H., Partin, L.R. (1984). Reactive distillation process for the production of methyl acetate. U.S. Patent 4,435,595.

[2] Siirola, J. J., (1996). Industrial applications of chemical process synthesis. Advances in Chemical Engineering 23, 1–62.

[3] Stankiewicz, A. I., Moulijn, J. A. (2000). Process Intensification: Transforming Chemical Engineering. Chemical Engineering Progress 1, 22–34.

[4] Lutze, P., Gani, R., Woodley, J. M. (2010). Process intensification: a perspective on process synthesis. Chemical Engineering Processing: Process Intensification 49 (6), 547–558.

[5] Reay, D., Ramshaw, C., Harvey, A. (2013). Process Intensification: Engineering for Efficiency, Sustainability and Flexibility. Butterworth-Heinemann, Waltham, MA.

[6] Babi, D. K., Cruz, M. S., Gani, R. (2016). Fundamentals of process intensification: a process systems engineering view. In: Segovia-Hernández, J.G., Bonilla-Petriciolet, A. (Eds.), Process Intensification in Chemical Engineering., http://dx.doi.org/10.1007/978-3-319-28392-0 2.

[7] Tula, A. K., Babi, D. K., Bottlaender, J., Eden, M. R., Gani, R. (2017). A computer-aided software-tool for sustainable process synthesis-intensification. *Computers & Chemical Engineering*. http://doi.org/10.1016/j.compchemeng.2017.01.001.

[8] Baldea, M. (2015). From process integration to process intensification. *Computers & Chemical Engineering*, *81*, 104-114.

[9] Demirel, S. E., Li, J., and Hasan, M. M. F., 2017. Systematic Process Intensification using Building Blocks, *Computers and Chemical Engineering*, http://dx.doi.org/10.1016/j.compchem eng.2017.01.04

[10] Papalexandri, K. P., Pistikopoulos, E. N. (1996). Generalized modular representation framework for process synthesis. *AIChE Journal*, *42*(4), 1010-1032.

[11] Ismail, S. R., Proios, P., Pistikopoulos, E. N. (2001). Modular synthesis framework for combined separation/reaction systems. *AIChE Journal*, *47*(3), 629-649.

[12] Iyer, S. S., Bajaj, I., Balasubramanian, P., Hasan, M. M. F. Modular Process Intensification of Carbon Capture and Conversion to Syngas. Submitted.

501702 Modeling and Simulation of a Hybrid Adsorptive-Membrane Reactor (HAMR) for Intensification of the Water-Gas Shift (WGS) Reaction Process

Abstract:

The Hybrid Adsorptive-Membrane Reactor (HAMR) is a dynamically operated process, which consists of a metal tube containing one or more tubular membranes. The HAMR's reaction section, surrounding the membrane tubes, contains catalyst and adsorbent material that accelerate reaction kinetics and adsorb some reaction products from the reacting mixture respectively. The HAMR's permeation section consists of the membrane tubes' interior, and physically removes some other reaction products from the reacting mixture. The HAMR is typically operated in reaction and regeneration modes, since the adsorbent meterial reaches its adsorption capacity after a period of operation, and thus needs to be regenerated.

In this work a mathematical model of HAMR is developed and simulated. The aim of this effort is to intensify the water gas shift (WGS) reaction process for hydrogen production. Indeed, the coupling of reaction and separation in the HAMR process can substantially improve reactant conversion, product selectivity, and can even yield near pure products. In addition, for reversible reactions such as the WGS, the HAMR process can be designed so as to deliver reactant conversions that exceed even equilibrium conversions attainable in conventional reactors. The HAMR system is composed of a, where some reaction products permeate. The proposed HAMR model quantifies velocity, species concentration, and temperature profiles throughout the reaction-adsorption, and permeation zones, by solving momentum/species/energy transport equations accounting for convection/reaction /diffusion/conduction mechanisms.

The developed model is used to intensify the Water Gas Shift Reactor (WGSR) Process. To this end, the performance of the traditional packed bed reactor is first quantified. Then parametric studies of the HAMR are carried out, so as to identify maximum intensification designs. These studies include a broad range of operating conditions and parameters (e.g. reactor operating temperature, catalyst and adsorbent weight to feed flowrate ratios, and others.

496925 Intensification of Chemical Processes Via Periodic Operation

Abstract:

Process intensification has emerged over the past few decades as a framework for increasing process performance by bringing chemical and physical phenomena into closer physical proximity. Intensification often transcends the traditional unit operations framework, with several conventional operations occurring in the same physical device. While this minimizes transport limitations and reduces equipment size, it can also lead to control challenges: intensified processes feature strong dynamic coupling between phenomena and have fewer control degrees of freedom than their conventional counterparts[1].

In this presentation, we describe a new means of operating intensified processes, focusing on deliberately imposing a transient (specifically, periodic) operating regime in order to improve average process performance. Additionally, we discuss periodic operation in terms of expanding the number of control degrees of freedom, in the sense that the amplitude and period of oscillation of a single manipulated variable can be set independently.

Using a prototype process featuring several coupled (and possibly competing) phenomena, we demonstrate theoretically that periodic operation can increase overall economic performance. We perform simulations to illustrate our theoretical findings, using both the canonical example described above, and the model of a dividing-wall column.

References

[1] Schembecker, G.; Tlatlik, S. Process Synthesis For Reactive Separations. Chem. Eng. Process. 2003, 42, 179-189.

498144 Intensification of the Hydride Vapor Phase Epitaxy Manufacturing Process for Solar Devices

Abstract:

Photovoltaics made of III-V compound semiconductors, especially in the form of multi-junctions, have demonstrated great conversion efficiencies [1]. However, their application is limited significantly by the expensive fabrication methods and materials. To address this challenge, hydride vapor phase epitaxy (HVPE) has regained research interest because of its intrinsically higher growth rate and lower precursor costs compared with other epitaxial growth methods. But the high growth rate and near-thermodynamic-equilibrium nature of this process also impose difficulties of producing large-area uniform wafer in a controlled way. The process needs to be geared into a more economic and manageable configuration to produce high quality products with stable properties at rapid throughput.

There are already a couple of practices applying the philosophy of process intensifications in this area. For example, through clever design, the compact horizontal reactor in [2] was capable of growing various III-V semiconductors (GaAs, InGaAs, GaInP, etc.) at high growth rates and depositing multilayers of those compounds in a controlled mode by manually operating transfer rod and shifting gas flow [2, 3]. Based on this reactor, an improved inline reactor was proposed [4], in which the growth method was adapted to sequentially deposit different layer of material on one wafer travelling through multiple reaction chambers separated by gas curtain. To further promote performance of process and product, we present novel alternatives to previous intensification techniques using new concepts in both equipment design and operation method. Computational modeling tools are used to verify design and operational strategies. We are expecting this work will not only aid the production of economically and industrially viable III-V solar cells, but also inspire more ideas to optimize other CVD processes that grow films with high cost-performance ratio.

[1] M. A. Green et al., "Solar cell efficiency tables (version 49)," Prog. Photovolt: Res. Appl., vol. 25, no. 1, pp. 3–13, Jan. 2017.
[2] K. L. Schulte, W. L. Rance, R. C. Reedy, A. J. Ptak, D. L. Young, and T. F. Kuech, "Controlled formation of GaAs pn junctions during hydride vapor phase epitaxy of GaAs," Journal of Crystal Growth, vol. 352, no. 1, pp. 253–257, Aug. 2012.
[3] K. L. Schulte et al., "Metalorganic vapor phase growth of quantum well structures on thick metamorphic buffer layers grown by hydride vapor phase epitaxy," Journal of Crystal Growth, vol. 370, pp. 293–298, May 2013.
[4] D. L. Young, A. J. Ptak, T. F. Kuech, K. Schulte, and J. D. Simon, "High throughput semiconductor deposition system," US20130309848 A1, 21-Nov-2013.

501295 Process Intensification of Reactive Separator Networks through the Ideas Conceptual Framework

Abstract:

A method to rigorously identify the performance limits of a reactive separator network is presented in this work. The quantification of the enhancement potential for a given technology can greatly benefit process intensification studies in the pursuit of radical improvements. The Infinite DimEnsionAl State-Space (IDEAS) conceptual framework is first reviewed and then shown to be capable of assessing the potential for process intensification of reactive separation processes. The IDEAS framework is employed to formulate an infinite linear program (ILP) that can synthesize optimal reactive separation networks, and establish rigorous tradeoffs between total network reactive holdup, and total network capacity. The proposed process intensification method is demonstrated on a case study involving the metathesis of 2-pentene through reactive distillation, an important chemical process for the oil industry. Total reactive holdup and total capacity, a surrogate for the amount of catalyst used and capital cost respectively, were selected as the systemâ€™s variables of interest for process intensification. Using a traditional reactor-followed-by-separation-system scheme as a baseline, an intensified IDEAS based reactive separation network can reduce the total reactive holdup and total capacity by 94.5% and 84.3% respectively. Results available in the literature were converted to the total holdup-total capacity space in the extent possible and compared with the IDEAS results. The comparison shows the existence of improvement potential, indicating that process intensification opportunities can be further explored in this reactive separation system.

490321 Intensified Reaction-Separation Schemes

Abstract:

Intensified reaction-separation schemes

Evangelia Koumaditi, John Woodley, Rafiqul Gani *Department of Chemical and Biochemical Engineering, Technical University of Denmark, DK-2800 Lyngby, Denmark*

The environmental and economic concerns are driving the chemical and biochemical industries to develop new, sustainable methods of production. Combining the reaction and the separation process in one unit, not only can it lead to cost and energy reductions for the overall process, but it can also have an effect on the reaction, in terms of yield. Motivated by the need to take advantage of the intensified process units, a method needs to be developed, which will allow the systematic generation and analysis of reactive separation configurations at an early stage in process synthesis. The objective of this work is the development of a method to design reaction-separation schemes, based on analysis of phase equilibrium of the involved chemical system.

Oxidation reactions are used as a generic system to highlight the conceptual basis of the design method. The selective oxidation of primary alcohols are examples of this generic system, which are used as case studies for the application of the design method. The properties of the pure components and the characteristics of phase equilibrium (azeotropes and liquid phase split) have been considered to identify new and novel intensified reaction-separation schemes. Also, the selective oxidation of alcohols to aldehydes is a challenging reaction, as aldehydes are relatively unstable products and also due to their use in the production of high-value products. Even though the reactions are considered to be bio-catalyzed by the enzyme Galactose Oxidase, the design method nevertheless is applicable to other similar systems and also not restricted to the use of biocatalysts only.

The specific examples highlighted in this work cover ternary systems involving alcohol + water + aldehyde, from which aldehyde is the product to be recovered, while alcohol and water are recycled back to the reactor. The first step of the method is the definition of the reaction system and data collection. The designs of the intensified separation schemes are based on the properties of pure components, as well as the properties of the mixture of the involved chemical system. The properties of pure components that are of interest are namely, the boiling point, the vapor pressure and the water solubility. After the pure component analysis is completed, the vapor-liquid and liquid-liquid equilibria are examined for each binary subsystem of the ternary mixture, employing available data or generated through model-based phase equilibrium predictions. The next step involves the classification of the systems according to their characteristics. The reaction-separation schemes are then determined by using phase equilibrium relations that favor the removal of the product from the reacting phase. Calculations of single-stage equilibrium based separations are performed in order to obtain the mass balance and the distribution of the compounds in the different phases. The last step involves validation tests to confirm the feasibility of the suggested configuration, by super-imposing the calculated mass balance lines on the predicted VLE and LLE plots. Success of the designed intensified reaction-separation scheme depends on the amount of the targeted compound removed from the reactor. Note that the separation and recovery of the product requires the use of solvents. Ionic liquids, because of their special properties have been considered as the needed solvents.

The presentation will highlight the results of the intensified reaction-separation schemes for a class of oxidation of primary alcohol reactions. Details of the design method together with the important concepts will also be highlighted.

499187 Process Intensification of Large-Scale Continuous Biobutanol Production via a Multi-Feed Bioreactor with *in situ* Gas Stripping

Abstract:

The economical production of alternative liquid fuels is vital to achieving energy independence from fossil fuel sources. While ethanol is often considered the primary candidate, butanol offers the advantages of higher energy density, lower vapor pressure, and lower volatility [1]. However, butanol is more toxic to cell cultures than ethanol, making it difficult to produce biologically at levels necessary for economic viability. Process intensification has been used in bioreactors by combining acetone, butanol, and ethanol (ABE) fermentation with separation phenomena, including pervaporation and gas stripping, to selectively remove the ABE products and overcome this limitation [2, 3]. Additionally, multi-feed inlet configurations have been investigated in continuous bioprocessing systems to limit inhibitory effects in bioreactors by manipulating the substrate concentration and dilution rate independently, limiting substrate concentration in the reactor without necessarily decreasing the reactor productivity [4]. The development of an intensified bioreactor with a multi-feed inlet configuration has the potential to overcome the toxicity limitation, increase process throughput, and produce butanol in a continuous, economically attractive manner that can compete with current bioethanol processes.

In this work, we examine the economic viability of large-scale, continuous production of biobutanol using *Clostridium acetobutylicum* and *in situ* gas stripping. First, the optimal control problem for a multi-feed intensified bioreactor is investigated using the flowrate of the stripping gas and inlet feeds as manipulated variable to maximize profitability. Batch reaction kinetics for the system, taken from Votruba et al., are extended to continuous processing in the formulation of the optimal control problem [5]. Solutions for the intensified bioreactor are compared to the economic viability of an optimally-controlled, continuous biobutanol reactor in the absence of gas stripping to emphasize the necessity of intensified process operations.

Once the optimal control policy for the bioreactor is determined, the process synthesis problem for large-scale biobutanol production from lignocellulosic biomass is explored by maximizing profitability. The synthesis problem considers using multiple bioreactors in parallel to process glucose made available from five options of cellulosic biomass, hybrid poplar, corn stover, sorghum, sugarcane bagasse, and switchgrass, in combination with four possible pretreatment methods, ammonia fiber expansion (AFEX), dilute sulfuric acid, sodium hydroxide, and liquid hot water. The use of lignin as a source of steam and salable electricity generation is also considered. The results of this study look to emphasize the importance of process intensification practices in the areas of bioprocess to develop new and otherwise economically impractical processes.

1. Liu, X.B., et al., *Enhancement of butanol tolerance and butanol yield in Clostridium acetobutylicum mutant NT642 obtained by nitrogen ion beam implantation.* Journal of Microbiology, 2012. **50**(6): p. 1024-1028.
2. Qureshi, N. and H.P. Blaschek, *Recovery of butanol from fermentation broth by gas stripping.* Renewable Energy, 2001. **22**(4): p. 557-564.
3. Li, J., et al., *Efficient production of acetone-butanol-ethanol (ABE) from cassava by a fermentation-pervaporation coupled process.* Bioresource Technology, 2014. **169**: p. 251-257.
4. Raftery, J.P., M.R. DeSessa, and M.N. Karim, *Economic improvement of continuous pharmaceutical production via the optimal control of a multifeed bioreactor.* Biotechnol Prog, 2017.
5. Votruba, J., B. Volesky, and L. Yerushalmi, *Mathematical-Model of a Batch Acetone Butanol Fermentation.* Biotechnology and Bioengineering, 1986. **28**(2): p. 247-255.

501636 Comparative Study of a Hybrid Adsorptive-Membrane Reactor (HAMR) with a Membrane Reactor/Adsorptive Reactor Sequence

Abstract:

Reactive separation processes such as membrane reactors (MRs) and absorptive reactors (ARs) have been attracting renewed interest for industrial applications. We have been studying a novel process that combines the MR and AR functions and which shows good promise for application in the efficient ultra-pure hydrogen production with simultaneous CO_2 recovery and capture.. Two system configurations are being investigated: One (known as the Hybrid Adsorptive-Membrane Reactor or HAMR) in which the membrane, adsorbent and catalyst are all housed in the same unit; and another in which the MR and the AR are separate but coupled units (known as the MR-AR system). The aim of this study is to perform a comparative numerical study between the HAMR and the combined MR-AR system, in which the MR is followed by an adsorptive reactor (AR), for the case-study of the water gas shift (WGS) reaction process for hydrogen production and carbon dioxide capture.

The HAMR is a dynamically-operated process, which typically employs one or multiple tubular membranes. The HAMR's reaction section, surrounding the membrane tubes, contains catalyst and adsorbent material that accelerates reaction kinetics via the adsorption of one (or several) of the reaction products. The HAMR's permeation section consists of the membrane tubes which physically remove some of the other reaction products from the reacting mixture, further accelerating reaction kinetics. The HAMR is, typically, operated in reaction and regeneration modes, since the adsorbent meterial reaches its adsorption capacity after a period of operation, and thus needs to be regenerated.

The MR-AR configuration (with the AR following the MR) provides significant added flexibility for the application, for which in addition to efficient hydrogen production CO_2 recovery and purity are also important key drivers. In the MR, a membrane that is selective to hydrogen is used to enhance the WGS reaction rate, and to potentially overcome equilibrium conversion limitations imposed by thermodynamics. The MR system is composed of a reaction zone packed with catalyst pellets, and a permeation zone, where the reaction products permeate. The AR contains both catalyst and adsorbent in a fixed-bed configuration for simultaneous reaction and separation. The combined MR-AR as with the HAMR is a dynamically operated process.

In the study, the velocity and species concentration profiles along the reactors' lengths are captured by momentum/species transport models accounting for convection/reaction /diffusion mechanisms. The model's equations are solved using Finite Element Method (FEM). The rigorous Maxwell-Stefan and dusty gas models are applied to describe mass diffusion fluxes. The developed model is used to intensify the Water Gas Shift Reactor (WGSR) Process. Then parametric studies of the HAMR and combined system are carried out, so as to identify maximum intensification designs. These studies include a broad range of operating conditions and parameters (e.g., reactor operating temperature, catalyst and adsorbent weight to feed flowrate ratios, and others).

493317 Zeolite Membrane Reactor for High-Temperature Isobutane Dehydrogenation Reaction: Effects of Membrane Properties and Operating Conditions

Abstract:

Zeolite Membrane Reactor for High-Temperature Isobutane Dehydrogenation Reaction:

Effects of Membrane Properties and Operating Conditions

*Shailesh Dangwal, Ruochen Liu, Seok-Jhin Kim**

School of Chemical Engineering, Oklahoma State University,

420 Engineering North, Stillwater, OK 74078, USA

ABSTRACT

Isobutylene plays an instrumental role as an intermediate in many important chemical engineering processes. Its use in making gasoline oxygenates, fuel additives and butyl rubber adds up to its demand and stimulates research for new energy efficient ways for its production. Catalytic dehydrogenation of isobutane using a membrane reactor is a feasible solution not only because it helps in overcoming thermodynamic limitations by shifting the equilibrium towards isobutylene but also it substantially separates isobutylene by preferentially allowing only H_2 to permeate through the membrane. In this study, dehydrogenation of isobutane in packed-bed membrane reactors operating with a Pt/Al_2O_3 catalyst is reported. The effects of MFI zeolite membranes on isobutane conversion, isobutylene selectivity and isobutylene yield are investigated experimentally and by simulations using a simple one-dimensional model. MFI membrane reactors allow the thermodynamic equilibrium limit of isobutane conversion to be surpassed at high temperatures. The experimental results and the model calculations demonstrate that medium-pore MFI membranes with good $H_2/i\text{-}C_4H_{10}$ selectivity can effectively enhance isobutane conversion at high operation temperature by timely removal of H_2 through the membranes. *Corresponding author: E-mail: seokjhin.kim@okstate.edu

499695 Ideas As a Process Intensification Tool Applied to Process Networks Containing Membrane Reactors

Abstract:

Process intensification is a strategy that can make dramatic reductions (order 100 or more) in the size of a chemical plant while meeting production objectives[1]. More recently, process intensification can be defined as "any chemical engineering development that leads to substantially smaller, cleaner, and more energy efficient technology[2]". In this work, the Infinite-DimEnsionAl State-space (IDEAS) conceptual framework is employed as a process intensification activity applied to process networks containing membrane reactors. The IDEAS framework enables a broad search of the design state-space to identify optimal process networks through the formulation of linear programs. This framework can assess fundamental performance limitations of networks of technologies under consideration in order to identify intensified designs. The IDEAS methodology for membrane containing networks will be applied to a case study featuring natural gas based hydrogen production, where energy efficiency will be intensified.

Membrane reactors are an emerging technology in the steam-methane reforming (SMR) process; these reactors typically involve simultaneous reaction and separation[3]. The biggest advantage of this technology is that it allows in situ separation, which gives rise to a high-purity product. This is particularly useful in driving the forward reaction of equilibrium limited reactions, as product is continuously removed from the reactor. The use of equilibrium reactor models allows for the search of intensified designs to be carried out in the lower-dimensional space of atomic fraction ratios, rather than the space of species mole fractions.

The process network considered will feature pressure changing devices, equilibrium reactors, a heat exchange network, flash separators, ideal separators, and membrane reactors. Operating and capital cost studies are all presented to determine the process network's cost relative to traditional steam methane reforming.

1. Ramshaw, C. The incentive for process intensification. BHR Group Conference Series Publication, vol. 18, pp. 1-4. Mechanical Engineering Publications Limited, 1995.
2. Stankiewicz AI, Moulijn JA. Process intensification: transforming chemical engineering. Chemical Engineering Progress. 2000;96(1):22-34.
3. Soltani, S., Sahimi, M., and Tsotsis T.T., "Catalytic Membrane Reactors: A Brief Overview", Encyclopedia of Membrane Science and Technology, Eric. M.V. Hoek and Volodymyr V. Talabar Editors, John Wiley & Sons, 2013.

500575 Additive Manufacturing for Process Intensification: Tailor-Made Design of Catalyst Supports for Single Phase and Multiphase Reaction Systems

Abstract:

Open cellular structures are promising novel catalyst supports as they combine the advantages of randomly packed fixed-bed reactors (radial mixing, tortuosity of the flow) and honeycombs (high geometric specific surface area, low pressure drop) owing to their high porosities and their characteristic 3D cellular architecture. Open cellular structures are manufactured mainly by means of a replication technique, which results in an irregular architecture of the structure, and consequently of the open porous system.

With the rise of additive manufacturing techniques it is nowadays possible to manufacture periodic open cellular structures (POCS) by, e.g., selective electron beam melting (SEBM) with well-defined geometrical properties. In this regard, a new degree of freedom in the design is unlocked as metal-based open cellular structures of nearly arbitrary geometrical complexity can be manufactured in a well-defined and highly reproducible manner.

We have extensively investigated open cellular materials with both periodic [1,2] and random [3] cell geometries regarding their performance as catalyst support. These structures were characterized with respect to their morphological/geometrical properties, fluid dynamic behavior and mass and heat transport properties. Models for the estimation of properties important for reactor design such as, e.g., specific surface area and pressure drop [1, 2] were developed and validated by experiments.

Detailed numerical studies of such geometries and the 3D flow field along with mass and heat transport processes as well as studies on the effective heat conductivity in POCS of different configurations were carried out [3, 4]. The established models allow for the design and optimization of POCS that are tailor-made according to the needs of the reaction system, representing a new class of superior catalyst supports.

The tailor-made POCS were manufactured, functionalized by catalytic coating, and then applied in two case studies for demonstration. In the first case study, the focus is on the optimization of heat transport in a highly exothermic gas phase reaction for enhancing the selectivity, where the oxidation of methanol to formaldehyde serves as example system. In the second case study the phase contacting, i.e. the gas-liquid distribution in a trickle-bed reactor was optimized [5]. As example reaction system, the hydrodesulfurization (HDS) of gas oil was investigated.

For both reaction systems, the tailor-made POCS were benchmarked against a traditional randomly packed bed reactor, and clearly a great potential for process intensification could be demonstrated.

References

[1] Inayat, A.; Schwerdtfeger, J.; Freund, H.; Koerner, C.; Singer, R.F.; Schwieger, W.; Chem. Eng. Sci. 66(12) (2011) 2758-2763.

[2] Klumpp, M.; Inayat, A.; Schwerdtfeger, J.; Koerner, C.; Singer, R.F.; Freund, H.; Schwieger, W.; Chem. Eng. J. 242 (2014) 364-378.

[3] Bianchi, E.; Groppi, G.; Schwieger, W.; Tronconi, E.; Freund, H.; Chem. Eng. J. 264 (2015) 268-279.

[4] Bianchi, E.; Schwieger, W.; Freund, H.; Adv. Eng. Mater. 18(4) (2016) 608-614.

[5] Laemmermann, M.; Schwieger, W.; Freund, H.; Catal. Today 273 (2016) 161-171.

485501 Light Absorption Efficiency of Gas-Liquid Continuous Flow Microreactors Illuminated By Visible LED Light Sources

Abstract:

The use of light energy constitutes a green activation mode of organic molecules. A photoreaction of increasing interest is the photosensitized addition of singlet oxygen, as it is an atom-economic method of functionalization and is applied in the synthesis of commercial chemical products as fragrances and pharmaceuticals [1].

Continuous photo microreactors are a promising tool for e.g. photooxidation transformations due to their small penetration depth and the presence of segmented flow which leads to enhanced mass transfer of the gas in the liquid [2]. In addition, the large photon flux through the thin liquid film located between the bubble and the reactor wall and light scattering at the gas-liquid interface were also reported as possible factors to enhance the photoreaction yield in comparison to single phase flow [3, 4]. However, designing an efficient photoreactor does not only involve optimizing its geometry, but also the light source needs to be adapted to the chosen chemistry, reactor channel layout and the phase of the reaction medium [5, 6]. In recent years, the use of LEDs has become popular in flow photochemistry due to their energy efficiency and the possibility to match their emitted wavelength to the photosensitizer absorption and their size to the reactor microchannels. However, the influence of the LED light distribution and potential scattering of light at the gas-liquid interface on the yield of the photochemical reactions performed in segmented flow is so far not reported.

This study aims to investigate the influence of the LED board configuration on the photon absorption efficiency in gas-liquid continuous flow microreactors. Firstly, the distribution and number of photons emitted by LED boards with different configurations is experimentally determined by near field goniophotometric measurements (NFGM). Secondly, the number of photons absorbed in the microreactor channels is quantified using an actinometer system. The balance of the photon fluxes for various LED light sources and for operating in single phase liquid or two-phase gas-liquid flow are compared and discussed. Furthermore, we developed a modelling tool based on the ray tracing technique to simulate the distribution of the scattered light rays from the gas-liquid interface, which supports the discussion of the experimental observations.

This study will contribute to the understanding of the incident radiation and dispersed phase influence on the photon absorption in gas-liquid photochemical transformations, knowledge which will be crucial for the efficient design of multiphase flow photochemical systems.

References

1. J. P. Knowles, L. D. Elliott, and K. I. Booker-Milburn, Flow photochemistry: Old light through new windows. Beilstein Journal of Organic Chemistry, 2012. 8: p. 2025-2052.

2. A. Gavriilidis, A. Constantinou, K. Hellgardt, K. K. (Mimi) Hii, G. J. Hutchings, G. L. Brett, S. Kuhn, and S. P. Marsden, Aerobic oxidations in flow: opportunities for the fine chemicals and pharmaceuticals industries, Reaction Chemistry & Engineering, 2016. 1: p. 595-612.

3. A. Yavorskyy, O. Shvydkiv, C. Limburg, K. Nolan, Y. M. C. Delauré, and M. Oelgemöller, Photooxygenations in a bubble column reactor, Green Chemistry, 2012. **14**: p. 888-892.

4. M. Nakano, Y. Nishiyama, H. Tanimoto, T. Morimoto, and K. Kakiuchi, Remarkable Improvement of Organic Photoreaction Efficiency in the Flow Microreactor by the Slug Flow Condition Using Water, Organic Process Research and Development, 2016. 20: p. 1626–1632.

5. M. E. Leblebici, G. D. Stefanidis, T. Van Gerven, Comparison of photocatalytic space-time yields of 12 reactor designs for wastewater treatment, Chemical Engineering and Processing, 2015. 97: p. 106-111.

6. D. Ziegenbalg, G. Kreisel, D. Weiß, and D. Kralisch, OLEDs as prospective light sources for microstructured photoreactors, Photochemical & Photobiological Sciences, 2014. 13: p. 1005-1015.

502576 Fast Mixing, Ketimine Additions, and the Application of Flow Chemistry to Verubecestat

Abstract:

Alzheimerâ€™s disease (AD) is an epidemic of ever increasing concern that affects over 24 million patients worldwide. Research suggests that the formation of Î²-amyloid plaque is central to the onset of AD and that inhibition of the amyloid pathway could lead to a disease modifying therapy. One of the first steps in the formation of plaque involves the enzyme Î²-secretase (BACE-1). Clinical programs for BACE inhibitors are progressing at a number of companies including MSD where Verubecestat (MK-8931) is in Phase III.

During the commercial route development of MK-8931, it was determined that the nucleophilic addition to an imine could be improved by performing the reaction in flow. The enhanced mixing lead to an increase in the yield of the reaction while excellent heat transfer properties of a flow reactor lead to the elimination of the need for cryogenic conditions.

The development of this process from an initial lab scale hit through a successful demonstration in the prep lab at multi kilogram scale to a pilot plant campaign on hundreds of kilograms is presented.

Synthesizing high selectivity bromobutyl rubber by a microreactor system

Pei Xie, Guangsheng Luo*, Kai Wang*, Peijian Wang

The State Key Laboratory of Chemical Engineering, Department of Chemical Engineering, Tsinghua University, Beijing, 100084, China

xie-p14@mails.tsinghua.edu.cn, 86-10-62797493

Sklo che

Outline

2

- **Background**
- **Experimental section**
- **Results and discussion**
- **Conclusions**

Bromobutyl rubber (BIIR)

High performance properties :

- Low-permeability
- Ageing resistance
- Ozone resistance

Δ High rautohension, mutual viscosity

Δ Excellent compatibility with other unsaturated rubber

Δ Fast vulcanizing speed

Wildly used in:

- Tubeless tyre
- Medical rubber products
- Rubber hoses
- …

Principle

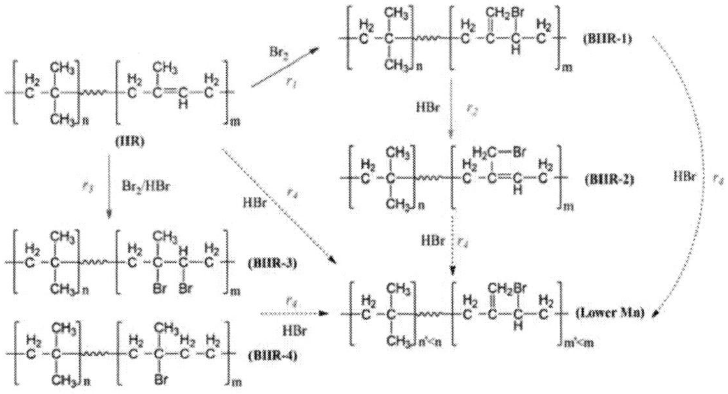

□ Characteristics of this system:

- High viscosity } Mixing-limited
- Extreme flow rate ratio of IIR solution to Br_2 solution } Mixing-limited
- Low selectivity of main product — Shorten resistence time & eliminate HBr quickly
- Strong corrosion from Br_2 and HBr — Develop a reactor made by corrosion resiatant material

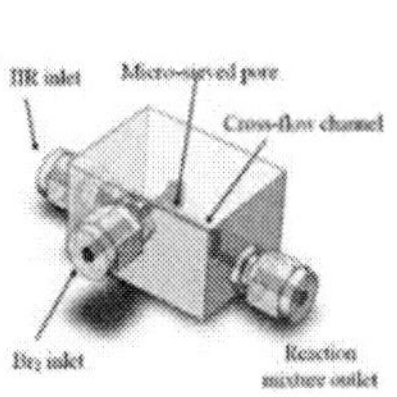

Experimental set-up 5

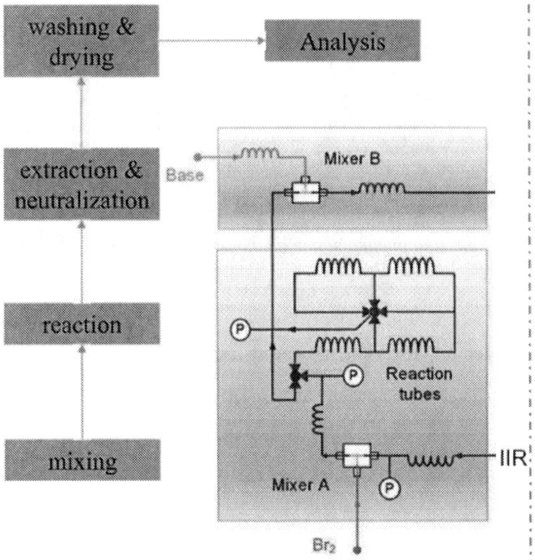

□ NMR analysis

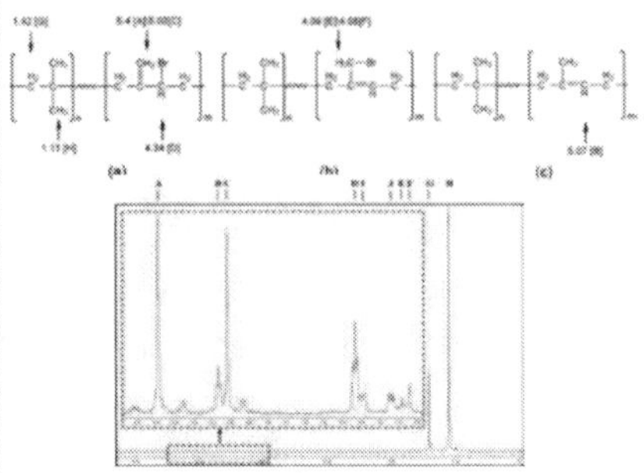

$$X = \frac{m_{cH\text{-}Br}}{m_{IIR}} = \frac{\left([D]+\frac{[E]+[F]}{2}\right)\cdot M_{Br}}{\left([D]+\frac{[E]+[F]}{2}\right)(M_{C5}-M_H+M_{Br})+[B]\cdot M_{C5}+\frac{[H]}{6}\times M_{C4}}$$

$$w = \frac{n_{BIIR\text{-}1}}{n_{BIIR\text{-}1}+n_{BIIR\text{-}2}} = \frac{[D]}{[D]+\frac{[E]+[F]}{2}}$$

$$U = \frac{\sum n_{\text{double-bond unit}}}{\sum n_{\text{double-bond unit}}+n_{C4}} = \frac{[B]+[D]+\frac{[E]+[F]}{2}}{[B]+[D]+\frac{[E]+[F]}{2}+\frac{[H]}{6}}$$

□ GPC analysis

- *Mn* (number average molecular weight)
- *PDI* (polydispersity index)

Outline 6

- **Background**
- **Experimental section**
- **Results and discussion**
- **Conclusions**

□ Influence of residence time

x_{IIR} (wt%)	x_{Br2} (wt%)	x_{NaOH} (wt%)	$x_{Na2S2O5}$ (wt%)
10.4	5.8	0.5	0.1
Q_{IIR} (mL/min)	Q_{Br2} (mL/min)	Q_{base} (mL/min)	T (°C)
90.3	9.4	49.7	25

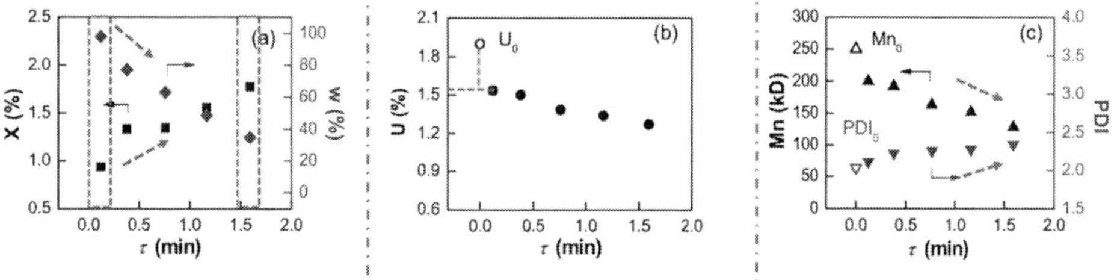

- HBr could catalyze the Br substitution position transfer reaction.
- Br_2 could also be consumed for addition reaction
- The additional rate: $r_{HBr} < r_{Br2}$
- HBr also plays a negative role in breaking the polymer chain

□ Importance of mixing intensification

x_{IIR} (wt%)	x_{Br2} (wt%)	x_{NaOH} (wt%)	$x_{Na2S2O5}$ (wt%)
10.4	6.2	0.5	0.1
Q_{IIR} (mL/min)	Q_{Br2} (mL/min)	τ (min)	T (°C)
23.8~100.2	2.4~9.9	1.44	25

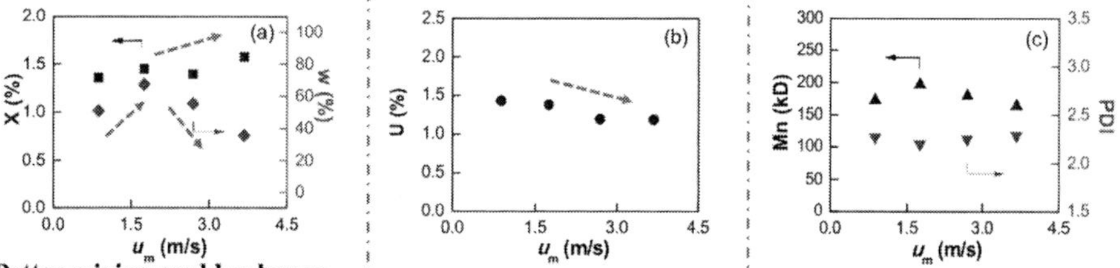

- Better mixing could enhance the conversion of substitution reaction
- U_m < 1.6 m/s, HBr cannot be extracted quickly by large droplets
- U_m > 1.6 m/s, quicker generation of HBr
- The enhancement of mixing also intensified Br_2 addition
- The polymer chains break seriously in this experiment

Results and discussion

□ Changing molar ratio of reactants

x_{IIR} (wt%)	x_{Br2} (wt%)	x_{NaOH} (wt%)	$x_{Na2S2O5}$ (wt%)
10.4	4.8~6.3	0.36~0.60	0.07~0.12
Q_{IIR} (mL/min)	Q_{Br2} (mL/min)	τ (min)	T (°C)
85.3	8.1~10.1	1.22~1.25	25

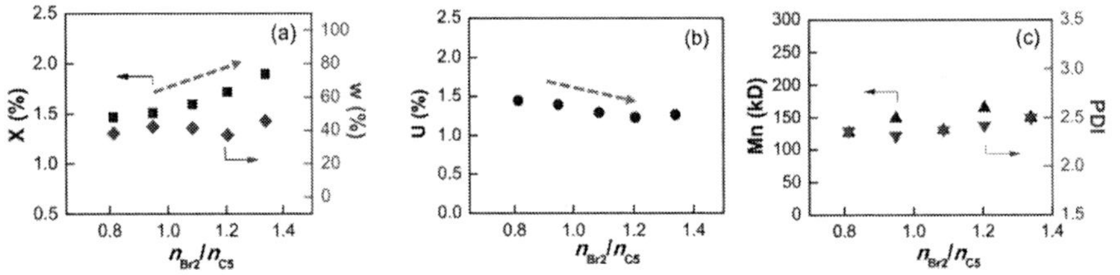

- **More Br_2 contributes to high mass concentration of substituted Br**
- **The excessive Br_2 also intensify the addition reaction**
- **High concentration of HBr keeps w and Mn at a low level**

Results and discussion

□ Effect of reaction temperature

x_{IIR} (wt%)	x_{Br2} (wt%)	x_{NaOH} (wt%)	$x_{Na2S2O5}$ (wt%)
10.5	6.0	0.5	0.1
Q_{IIR} (mL/min)	Q_{Br2} (mL/min)	τ (min)	T (°C)
95.3	9.7	1.51	25

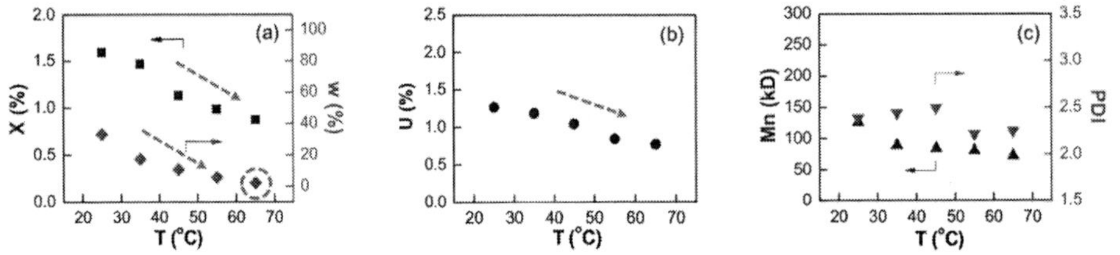

- **The high temperature accelerated the Br substitution position transfer reaction and the addition reaction**
- **The polymer decomposition was also easier at high temperature**

Results and discussion 11

☐ Water assistant method

x_{IIR} (wt%)	x_{Br2} (wt%)	x_{NaOH} (wt%)	$x_{Na2S2O5}$ (wt%)
9.9	6.3	0.5	0.1
Q_{IIR} (mL/min)	Q_{Br2} (mL/min)	x_{H2O} (wt%)	T (°C)
70~99.9	7.1~10.1	1.0	25

Pre-dispersed water droplets

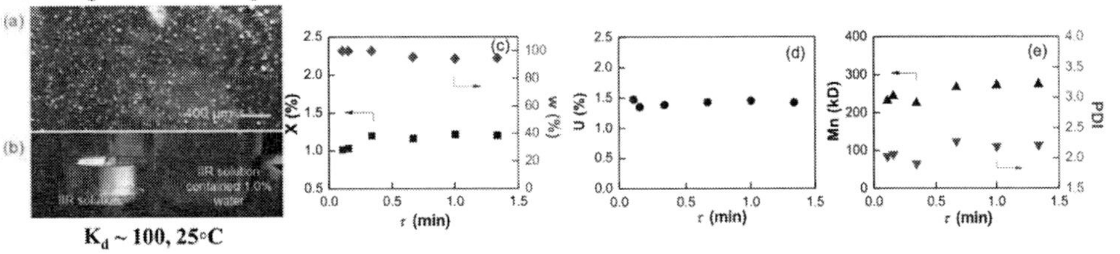

K_d ~ 100, 25°C

- The molar ratios of BIIR-1 are close to 100% even at high residence time
- More Br_2 would be extracted at long residence time with more HBr in the water phase
- This water assistant method works well for preventing the addition reaction

Outline 12

- **Background**
- **Experimental section**
- **Results and discussion**
- **Conclusions**

Conclusions

- The microreactor system, which had significant advantages in reactant mixing, less than a minute residence time control and low-cost corrosion prevention, worked well for synthesizing BIIR.
- Low reaction temperature was found a good way to prohibit addition reaction.
- A water assistant technology was proposed to prevent the Br transfer reaction and the polymer chain decomposition by extracting HBr in-time from the reacting mixture.

Thank you for your listening!

Any question will be highly appreciated!

500549 Modeling and Optimization of a Two-Dimensional Tubular Reactor for an Exothermic Reaction

Abstract:

As the world searches more energy resources at more remote locations, industry has been trying to optimize topside system of the offshore plant. One of the important aspects for this is an optimization of reactor in floating production storage and offloading (FPSO) system, where space and weight savings are crucial. In this study, a tubular reactor equipped with a cooling jacket for the exothermic reaction in topside platform was modeled based on the reaction mechanisms and kinetics obtained from experimental results. The influence of various design and operating variables (e.g. reactor length, gas flow rate, inlet temperature and concentration, and thermal conductivity) on conversion and temperature profiles along the axial and radial directions inside a reactor were studied. For this, the modeling was described in an axisymmetric coordinate system with finite element approximations using computer simulations. Lastly, the mathematical modeling of the reactor design was optimized to maximize an overall yield of the main products.

Acknowledgement
This work was supported by a research project of â€œDevelopment of Integrated Interactive Model for Subsea and Topside System to Evaluate the Process Design of Offshore Platformâ€, funded by Ministry of Trade (project No. 10060099).

483641 Liquid-Liquid Extraction: A Simplified View of a Complex Process

Abstract:

The Chemical Processing Industry (CPI) uses liquid-liquid extraction when distillation and rectification are difficult or ineffective. The CPI also uses liquid-liquid extraction to recover temperature-sensitive substances, substances that decompose or form undesired by-products at elevated temperatures. Liquid-liquid extraction is also a common unit operation in petroleum refining. For example, most refineries operate an Edeleanu process, a furfural process, a Duo-Sol process, a phenol process, or a propane process. All these processes involve a counter-current flow of solvent and distilled crude oil through a vertical column. However, some processes use a horizontal extraction vessel.

All liquid-liquid extraction processes are designed based on attainment of phase equilibrium. However, attaining equilibrium in a commercial process is challenging. Why? Because attaining equilibrium requires an infinite time period, which is never available to a commercial process.

This caveat with regard to equilibrium brings into question the procedure for designing a liquid-liquid extraction column. The current method involves using a triangular equilibrium plot, then projecting operating lines through an externally located focal point. The number of lines passing through the external focal point determines the number of stages required to reach a desired combination of raffinate-extract concentrations. In general, this number is not a whole number, which necessitates shifting the focal point empirically to obtain a whole number of lines. The size of the extraction column is then determined using the final number of lines. In summary, liquid-liquid extraction columns are built on an assumption never met industrially, then adjusting that assumption empirically to arrive at an acceptable result.

This presentation applies dimensional analysis to liquid-liquid extraction, which avoids the equilibrium assumption. In this presentation, we derive the pertinent dimensionless parameters necessary to describe a liquid-liquid extraction process and show how to use them to scale from one unit size to another unit size.

503040 The inside Story of an Extractive Distillation Column: Process Modelling Based Scale-up with Plant Data in a Debottlenecking Case Study

Abstract:

Distillation is the primary separation process used in petrochemical and refining industry. Sometimes it is technically difficult to separate close boiling compounds using normal distillation and it is economically expensive as well. Some of these close boiling compounds can be separated using extractive or reactive distillation columns. Extractive distillation, reactive distillation plays an important role by combining two unit operations in a single equipment to reduce CAPEX and OPEX.

In extractive distillation, a solvent is added to alter the boiling points of the compounds to be separated. Modelling of these advanced distillation processes is very important to optimize the operation as well as to retrofit and/or debottleneck the existing plants. The interaction between solvent and the components in the original mixture plays an important role in the extractive distillation. Thermodynamic vapour liquid equilibrium (VLE) models are used to describe these interactions. These VLE models use binary interaction parametes (BIPâ€™s) to quantify these interactions that will affect the boiling points of the compounds.

This paper explores the methodology of process modeling based scale-up by regressing the binary interaction parameters from laboratory as well as plant data. By using these parameters, the accuracy of the plant model was improved further and it is used for the debottlenecking study of a petrochemical plant.

492374 Applying CFD Towards Defining Crystallisation Scale up Strategy

Abstract:

Efficient scale up of a crystallisation is an objective that presents significant challenges for pharmaceutical industry. It is critical that particle characteristics remain robust throughout scale up from laboratory to plant.

Within this presentation, an examination of the use of novel technologies such as Computational Fluid Dynamics (CFD) in aiding the design of an experimental scale up strategy will be outlined. The standard approach within industry uses established engineering calculations in conjunction with experimental data obtained during lab scale experiments to determine scale up effects. This presentation will assess this strategy for a model API system across different scales and describe use of CFD as a complimentary tool. The use of CFD will be discussed in terms of the qualitative and quantitative aspects to the model outputs, such as directional velocities, torque and shear values, as well as its ability to predict or explain any impact to crystallisation output. The role of CFD in the design of investigative scale up experiments will be highlighted as well as the potential CFD offers to test vessel configurations before being implemented on plant.

498751 Lab Scale Experimental Workflow to De-Risk Production Equipment Change for Filtration and Drying of Active Pharmaceutical Ingredients

Abstract:

As an Active Pharmaceutical Ingredient (API) moves through development, the scale and site of manufacture is subject to change. At AstraZeneca we see significant production scale equipment changes between our pilot plant and commercial scale facilities. Particularly for the API stage, a change in the filtration and drying equipment used at production scale can result in changes to the quality attributes of the API. These quality attributes include particle size and shape, and the range of acceptability on secondary processing is often difficult to quantify. Therefore the preference is to maintain the particle properties within the range we have manufactured before. However, this leads to uncertainty whether we can achieve this goal when making substantial changes to the filtration and drying equipment, for example moving from an agitated filter dryer to centrifuge and spherical dryer.

One approach to de-risk equipment change has been to manufacture large scale trial batches using the new equipment and use the material in formulation manufacture to check that quality attributes are met and no operability issues are encountered. However, this requires the use of expensive intermediates, often with restricted availability, as well as requiring significant manufacturing time and effort in both the API and formulation manufacturing plants.

There is a range of filtration and drying equipment used at production scale and it would not be practical to have a scaled down version of each type of filter and dryer in our pilot plant.

Another significant consideration is that once the large scale trial has been performed, timelines are often too short to make an equipment change based on the results as the accommodation process for that compound into the particular plant has already begun.

Therefore the first step is to identify the key changes in the characteristics of the type of equipment being used. For example when moving from a pressure filter to a centrifuge the key change would be the increase in applied pressure across the cake due to the centrifugal force, potentially leading to cake compression and lump formation. Or, when moving from agitated filter dryer to spherical dryer, the type and duration of agitation is changed which could lead to agglomeration or attrition.

An experimental workflow has been designed to assess the impact of different filtration and drying equipment types at lab scale, using 200-500g of API. The material can then be blended with excipients using a formulation process such as roller compaction or wet granulation, followed by compaction using a tablet simulator. This allows characterisation of the API, the blend and final product form and any changes to product quality attributes can be identified.

One example of the use of this workflow is an API that had previously been isolated on an agitated filter dryer. The API was known to be prone to attrition so was minimally agitated during drying in the pilot plant. The manufacture of this API was then transferred to a commercial facility where a pressure filter followed by spherical dryer would be used so there was concern that significant attrition would be observed and therefore the particle size would be smaller than any of the previously manufactured batches. It was also not known if agglomeration would be observed.

We do not have a lab scale spherical dryer, but the key characteristics of rapid drying with continuous agitation were tested in a lab scale agitated filter dryer and also a conical dryer. It was found that some attrition was observed during the drying stage, and this was broadly comparable between the continuously agitated pressure filter and the conical dryer. Despite agitating for an extended period of time, the particle size did not drop below the size of smallest batch manufactured so far. It was also noted that there was a particle breakage in the pressure filtration - there was a decrease in particle size between the sample taken from the crystallised slurry and a sample taken from the wet cake. The material was then filled into capsules and dissolution testing performed. This experiment allowed us to show that the quality attributes of the API and product were maintained despite changing the drying equipment type. Samples taken from the manufacturing campaign showed that the particle size decreased during pressure filtration and again during spherical drying and the particle size of the dried material was within the range previously observed. This experimental protocol was repeated for a second API but no attrition was observed at lab scale and this was matched by observations at commercial scale, again in the spherical dryer.

This approach is currently being embedded into late stage development of API processes and will lead to more robust technical transfers of processes to our commercial facilities.

499780 Development of a Pegylated Protein for Large Scale Manufacturing

Abstract:

Polyethylene glycol (PEG) has been shown to have several applications in biotechnology, including the ability to improve molecule stability, solubility and extension of half-life through pegylation. By-product formation during the PEGylation process can be challenging for downstream purification to ensure product purity. This presentation will discuss the variables and optimization, for a protein that is PEGylated, in order to reduce the level of impurities and increase pegylation efficiency.

502125 Multiscale Approach to the Design of a Trickle Bed Reactor

Abstract:

This talk will discuss the scale-up of a trickle bed reactor for selective liquid phase hydrogenation. Insights from bench scale experiments, cold flow studies, pilot plant testing, and fundamentals based kinetics and reactor modeling are combined to increase the robustness and confidence in commercial reactor design.

491211 Effect of Non-Ideal Behavior on the Energy Minimum Design of Highly Integrated Reaction and Separation Processes

Abstract:

Process intensification has become an important concept to meet the challenges of an increasingly global competition in the chemical industry. A promising possibility of process intensification is the integration of several primarily separated process operations into one single shell. Examples of this are the dividing wall column or the reactive distillation, which show significant savings in investment and operational costs. Further savings can be achieved by an additional integration step of these two apparatus into the reactive dividing wall column (RDWC). Though, due to the high grade of integration the process behavior of all three of these integrated processes is strongly nonlinear. Hence, it is not trivial to predict their advantageousness for a given task during conceptual design.

However, in the industrial praxis it is essential to know, how the process performances and for which tasks the RDWC is more advantageous than less integrated process alternatives. Therefore, the aim of this research is to derive systematically conceptual design heuristics and generate a deep understanding of the process behavior. So the optimal process integration level can be determined easily and the conceptual design phase can be accelerated. The investigations carried out focus on the influence of non-ideal system properties, such as azeotropes and enthalpies. Therefore, various chemical systems with systematically changed properties are investigated. To determine the best process integration level, the energy optimal designs of the RWDC and less integrated process alternatives are determined and then compared. The application of this method allows the development of heuristics regarding the optimal process integration. The basic requirement for these investigations is a flexible simulation model which can describe the RDWC as well as less integrated process alternatives.

The presentation will not only show the overall concept of the research, but also focuses on the developed simulation model with its reliable convergence strategy. Additionally, the influence of non-ideal system properties on the process behavior will be presented and the energy minimum design of the RDWC compared to less integrated process alternatives will be demonstrated.

498711 Characterization and Evaluation of Zero-Length Covalent Crosslinking Strategies for DNA Based Applications

Abstract:

An adapted strategy for covalently coupling amine and phosphate functional groups in biological molecules was developed from the conventional 1-ethyl-3-(3-dimethylaminopropyl)carbodiimide hydrochloride (EDC) crosslinking method. The coupling mechanism was studied with synthetic single stranded DNA (ssDNA) and ethylenediamine (EDA). The study demonstrates that the phosphoramidated ssDNA conjugate is stable for up to 14 days upon storage at -20 Â°C. It also provides evidence for stable covalent phosphoramidation exclusively at the 5â€™ phosphate group of the ssDNA. Fourier transform infrared spectroscopy (FTIR) was utilized for bond characterization of the conjugates. The extent of reactions were determined by reversed phase liquid chromatographic methods (RP-HPLC). A comprehensive study on analytical method validation was conducted to determine parameters such as the accuracy and detection limits of the developed HPLC analytical method. Finally, we report the yield of phosphoramidated products synthesized through the two separate crosslinking strategies, and provide a conclusive comparison between the two reaction approaches.

The current study on crosslinking strategies using EDC and imidazole is presented as a convenient one-pot reaction method for areas of research where there is a great lack of flexibility in altering reaction conditions such as the temperature or pH, to obtain higher product yields. It is directly applicable to studies in biomedical diagnostics, biosensor development, and self-assembled biomaterial synthesis.

495381 Adsorption Desulfurization Performance of B_2O_3 Modified Ag-CeO_x/TiO_2-SiO_2 and Adsorption Diffusion Study

Abstract:

Adsorption desulfurization performance of B_2O_3 modified Ag-CeO_x/TiO_2-SiO_2 and adsorption diffusion study

Meiqin Zheng[1], Xiaohui Chen[1*]

[1]National Engineering Research Center of Chemical Fertilizer Catalyst, School of Chemical Engineering, Fuzhou University

* Correspondence author E-mail: chenxhfzu@fzu.edu.cn.

In this word, the effect of B_2O_3 modified Ag-CeO_x/TiO_2-SiO_2 adsorbent on adsorption desulfurization and adsorption diffusion behaviour was investigated. The adsorption desulfurization performance was studied by static batch and fixed bed test through CN-IV diesel (sulfur content 32.78 mg/L). The pore diffusion model (PDM) and homogeneous surface diffusion model (HSDM)[1] were employed to investigate the adsorption diffusion behaviour of 4,6-dimethyldibenzothiophene (4,6-DMDBT) on model diesel. Simultaneously, a 2D porous media CFD model is developed for a fixed bed adsorber to remove the 4,6-dimethyldibenzothiophene (4,6-DMDBT) from diesel. The batch tests proved that the sulfur concentration after desulfurization was less than 10 mg/L, which can meet the CN-IV standard.

The B_2O_3 modified CeO_x/TiO_2-SiO_2 support was prepared by co-impregnation method with 1:1:2:44.3 molar ratio of B:Ce:Ti:Si. Silver was dispersed onto the B_2O_3 modified CeO_x/TiO_2-SiO_2 support through typical incipient wetness method[2]. The adsorbents were characterized by means of X-ray diffraction, N_2-adsorption specific surface area measurements (BET), images of transmission electron microscopy (TEM), high-resolution transmission electron microscopy (HRTEM) and high angle annular dark field scanning transmission electron microscopy (HAADF-STEM) etc. The results proved that B species was dispersed evenly on the supports, and didnâ€™t change the pore structure of adsorbents. It was found that B_2O_3 species exhibited a great enhancement for the adsorption desulfurization activity because of a significant enhancement for the amounts of weak acid sites generated from more BO_4 species on surface of the adsorbent[3,4].

The batch equilibrium and kinetics of adsorption desulfurization of model diesel fuel were also investigated. The proposed HSDM model and PDM model can successfully describe the adsorption of 4,6-DMDBT for different operating conditions studied. The adsorbate was absorbed rapidly initially because the adsorption rate at the beginning was dominated only by external diffusion. The Biot numbers proved that the predominant rate controlling step was intraparticle diffusion. Simultaneously, the adsorption rate was mainly controlled by surface diffusion at lower initial bulk phase concentration and less adsorbent dosage, while pore diffusion played a significant role at higher initial bulk phase concentration and more adsorbent dosage.

The CFD simulation results demonstrated that porous media CFD model can describe successfully the adsorption mass transfer process for removal of 4,6-DMDBT from diesel. The mass transfer rate increased with increasing feed concentration at the beginning due to the higher driving force at higher feed concentration, conversely, the lower mass transfer rate was obtained at higher feed concentration after a period of adsorption time because of the decrease in the saturation time at the fixed bed column entrance gradually at higher feed concentration.

Key words: adsorption desulfurization; B_2O_3; HSDM; PDM; CFD.

REFERENCES

1. Zheng M, Xu C, Hu H, Ye Z, Chen X. A modified homogeneous surface diffusion model for the fixed-bed adsorption of 4,6-DMDBT on Agâ€"CeO_x/TiO_2â€"SiO_2. *RSC Adv.* 2016;6:112899-112907.

2. Xu CZ, Zheng MQ, Chen K, Hu H, Chen XH. CeO_x Doping on a TiO_2-SiO_2 Supporter Enhances Ag Based Adsorptive Desulfurization for Diesel. *J. Fuel. Chem. Techno.* 2016;44(8):943-953.

3. Li-Da LI, Cheng-Zhi XU, Zheng MQ, Chen XH. Effect of B_2O_3 modified Ag/TiO_2 -Al_2O_3 adsorbents on the adsorption desulfurization of diesel. *J. Fuel. Chem. Techno.* 2015;43(8):990-997.

4. Peil KP, Galya LG, Marcelin G. Acid and Catalytic Properties of Nonstoichiometric Aluminum Borates. *J. Catal.* 1989;115(2):441-451.

503642 Experimental Vapor-Liquid Equilibrium Data for the Ternary Mixture (Methane + Propane + Methylbenzene) at Conditions Relevant to the LNG Scrub Column

Abstract:

LNG distillation (scrub) columns are used to remove heavy components that have the potential to freeze out, affecting the different operational units used during the liquefaction of natural gas (LNG) (e.g cryogenic heat exchanger (MCHE)). Such blockages are made of benzene, toluene, ethylbenzene and xylenes compounds, or BTEX. In order to avoid them, one can predict its occurrence by improving the existing models, in particular, cubic Equations of State (EOS). These models are based on Vapor-Liquid Equilibrium (VLE) experimental data and therefore itâ€™s accuracy is highly dependent on the data available.

In order to improve the efficiency and certainty, the VLE of a ternary methane + propane + methylbenzene (toluene) mixture was studied for different conditions, toluene being the minor component in both liquid and vapour phases. These measurements of ternary mixture were conducted within different isochoric paths at temperatures between 213 and 298 Kelvin and pressures up to 8.3 MPa. Two devices were used for the VLE measurements; the first device was used for measurements from (263 to 298) K while the second was mainly used for measurements down to 213Â K. Both apparatusâ€™ have a sample of the liquid and gas phases by means of using sampling valves and a GC that was fitted either with a FID or BID detector.

The HYSYS Peng Robinson (PR) EOS that is used widely in the LNG industry was used to compare it to the VLE data results. The concentration of toluene in the vapour phase was found to be under-predicted by the HYSYS PR EOS by an average of around 50 %, with the error increasing as temperature and toluene concentration decreased. These VLE measurements unambiguously demonstrate that the HYSYS PR EOS as well as other cubic EOSs substantially under-predict the possible toluene content of saturated vapours that could be present in the overhead of the LNG scrub column. This research highlights the importance of improving predictions of the allowable threshold concentration of BTEX in fluids entering the MCHE to avoid potential freeze-out events.

491152 Reactive Dividing Wall Column: Experimental and Simulative Studies about Process Control and Dynamic Behaviour

Abstract:

Process integration is an important method to increase process efficiency in chemical industry. A highly integrated process is represented by the reactive dividing wall column (RDWC), which combines a chemical reaction and several separation steps in one column shell. This integration can lead to major savings in investment costs and energy demand in comparison to conventional reaction and separation sequences. However, the high degree of integration also causes a very complex process behaviour. Hence chemical industry is reluctant to build and operate RDWCs. In order to assure constant product purities and a stable continuous operation a good understanding of dynamic effects as well as reliable process control strategies are required. To date various papers have been published about design and steady state operation of RDWCs but there are only few papers about the control and dynamic operation. Experimental studies about the control of RDWCs are still pending.

Our work aims to contribute to the research about RDWC process control. In our research we systematically develop and examine decentralized process control concepts for the RDWC. For the first time we conduct comprehensive experimental studies with focus on the dynamic behaviour and the control of RDWCs with an enzymatically catalysed RDWC in pilot scale. By comparison of data obtained experimentally and by simulation studies (all employed simulation models were developed at the Institute of Process and Plant Engineering, Hamburg University of Technology, and provide high reliability due to extensive experimental validation [1]) we are able to validate the developed control structures. Our studies allow deep insights about important dynamic effects of the apparatus and provide thus a basis for stable and safe process operation.

In the poster a suited decentralized process control strategy is shown, analysed and evaluated. Based on steady state simulations an optimal structure is established, while dynamic simulations are employed to adjust the controllers. Afterwards the control structure is tested in simulation studies in order to analyse the dynamic behaviour in different disturbance scenarios. In the next step the control structures are verified in experimental tests on our RDWC in pilot scale. Therefore, experimental data of the RDWC pilot plant will be shown. The dynamic behaviour under the influence of disturbances will be analysed in detail. Based on the results further developments for the control structure will be derived.

---

[1] Egger, T.; Fieg, G. (2016): Enzymatic Catalyzed Reactive Dividing Wall Column. Experiments and Model Validation. In *AIChE J. DOI:* 10.1002/aic.15598.

499058 A Thermodynamic-Based Modeling and Analysis Approach for Mechanical Energy Recovery

Abstract:

Process energy recovery has always been a major practice in process system engineering, as despite a tremendous effort, energy loss in manufacturing sectors is still very high (~50% of the total consumption) (Energetics Inc., 2014). Heat-based thermal energy and work-based mechanical energy are two common forms of energy that are being consumed significantly in the process industry. Note that mechanical energy receives increasing attention due to significant loss in process systems.

In this work, the concept and fundamentals of mechanical energy recovery using direct type of work exchangers will be briefly discussed. A thermodynamic analysis method that can be used to quantitatively identify the maximum amount of recoverable mechanical energy of process systems will be investigated, and its feasibility will be shown through isothermal and non-isothermal conditions. A comprehensive methodology for work exchange network synthesis will be introduced and the feasibility of the combined work and heat exchanger network synthesis design will be discussed. The integration methodology will be tested through deriving solutions for a case study with consideration of the economic feasibility of the system.

Reference:

Energetics Inc. Manufacturing energy and carbon footprint. Prepared for the U.S. Department of Energy, Advanced Manufacturing Office. 2014.

Influence of transport properties and correlations between properties in process modeling

Pia Herrmann[1], *Kevin Busch*[1], *Karsten Müller*[1,*]

[1] *Institute of Separation Science and Technology, Friedrich-Alexander-Universität Erlangen-Nürnberg (FAU), Germany*

* *To whom correspondence should be addressed: Karsten Müller (karsten.mueller@fau.de)*

Introduction

Simulation is an important tool for accessing efficiency of processes and can be used for identifying potentials for improving existing technologies. For example, molecular modeling can be used to design better working fluids for processes like the Organic Rankine Cycle[1], CO_2 capture[2] or heat pumps[3].

It has been demonstrated in previous works[4] how process simulation can be used to quantify the potential for improvement related to different parameters such as substance properties or improved apparatuses. For this approach it is important to know the sensitivity of the target property, that should be improved (e.g. efficiency), on the individual input parameters. The approach evaluates the improvement that realistically seems possible by finding e.g. working fluids with better properties. The potentials are assigned to the individual properties. Based on these results it can be seen if it is reasonable to spent research efforts on projects targeting parameters with small potential for actually improving the process. This enables making a decision on research project at an early stage and avoiding inefficient use of research resources.

This contribution aims at answering two questions relevant in this context, which have not yet been discussed in the previous works. First, the validity of the assumption that the input parameters can be varied independently is evaluated. Second, it is discussed how the potential related to finding working fluids with better transport properties can be accessed.

Correlations between substance properties

On the first glance, independent variation of input parameters, which are often substance properties, does not seem reasonable. However, due to the large number of known (and still unknown) chemical substances it is possible to find nearly every combination of substance properties. Nevertheless, some restrictions remain. For example, one can assume that a substance with a large molecule will have a low vapor pressure, but a large molar heat capacity. Hence, a negative correlation between these parameters can be expected and independent variation seems questionable.

To analyze the dependence between substance properties correlation coefficients have been evaluated. The Pearson correlation coefficients between six exemplary substance properties are shown in Table 1 (all values on the diagonal are unity, because every property has a perfect linear correlation with itself). As expected a negative correlation coefficient is observed for vapor pressure and heat capacity. However, most other properties correlate only very weakly. In most cases the correlation coefficients are close to zero, indicating no significant correlation. The potentials for process improvement related to these properties can be evaluated by independent variation. If correlation is stronger it can be advisable to consider a linear property-property-relationship. However, closer evaluation shows that the effect of intercorrelation is still negligible.

Table 1: Correlation coefficients between selected substance properties (evaluated on 200 organics)

	c_P	P^{LV}	λ	η	M	V_{molar}
c_P	1	-0.42	0.11	0.10	0.80	0.94
P^{LV}		1	-0.20	-0.09	-0.27	-0.38
λ			1	0.28	-0.72	-0.08
η				1	-0.01	-0.04
M					1	0.57
V_{mola}						1

Transport properties

In a second part the influence of transport properties such as viscosity and thermal conductivity on stationary process models is studied. Sensitivity on e.g. viscosity is often quite small. However, the effects related to viscosity can be pronounced, since viscosity can be varied within a large range of values. effects are evaluated in this study.

The relevance of transport properties should be demonstrated on the example of heat transfer. Heat transfer is an important aspect of many energy processes, because it determines e.g. the temperature difference in heat exchangers. This temperature dependence is determined by the heat transfer coefficient α. The heat transfer coefficient can be calculated using Nusselt correlations. Nusselt number is depending on Reynolds and Prandtl number and these again are determined by the substance properties viscosity η, density ρ, thermal conductivity λ and heat capacity c_P. Based on the Nusselt correlation it is possible to describe the dependence of heat transfer coefficient on these properties:

$$\alpha \sim \eta^{n_\eta} \qquad (1)$$

$$\alpha \sim \eta^{n_\rho} \qquad (2)$$

$$\alpha \sim \eta^{n_\lambda} \qquad (3)$$

$$\alpha \sim \eta^{n_{c_P}} \qquad (4)$$

The exponents n for equations 1 to 4 are given in Table 2. As expected, the exponent for viscosity is negative, because high viscosity inhibits convective heat transport. The effect is more pronounced in the turbulent regime, since convective heat transport is more important than conduction in turbulent regime. The opposite effect can be observed for thermal conductivity. As expected the exponents are positive and higher in the laminar regime.

Table 2: Exponents n for proportionality in heat transfer for viscosity, density, thermal conductivity and heat capacity

	η	ρ	λ	c_P
laminar	-0.17	0.50	0.67	0.33
transitional regime	-0.41	0.82	0.59	0.41
turbulent	-0.35	0.85	0.56	0.49

Using these correlations it is possible to describe the dependence of heat transfer (and thus temperature difference in heat exchangers). Considering the realistic variation range of the properties, the potential for improvement can be estimated (see Figure 1 for an analysis on the example of a LOHC based energy storage system). It turns out that viscosity has the strongest influence among the transport properties, even though the sensitivity is weakest. This can be explained by the fact that viscosity can be varied within a very large range of values. However, it also strikes that the potential for improvement related to transport properties is limited. Searching for better LOHC materials should therefore focus on substances with better (in that case less negative enthalpies of reaction for the hydrogenation reaction).

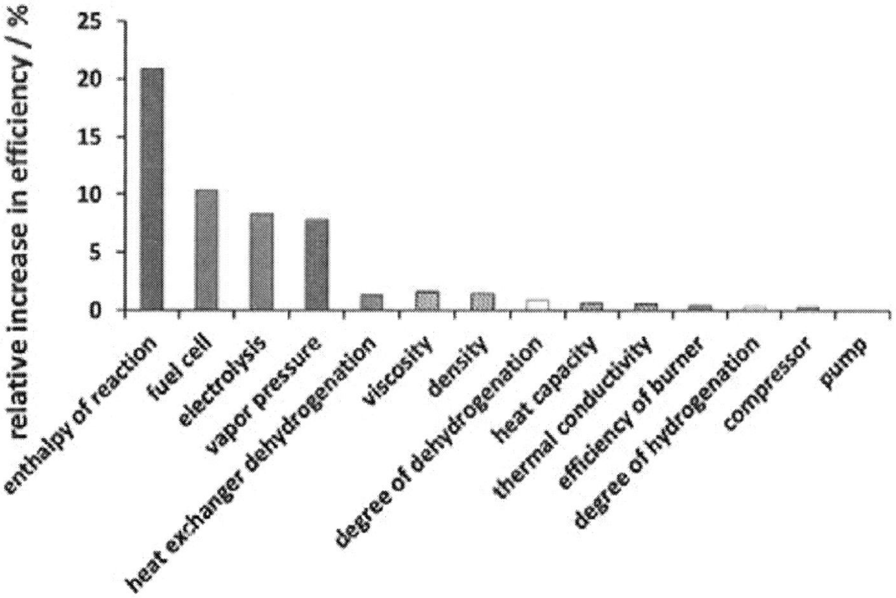

Figure 1: Potential for improvement in the LOHC process (red: substance properties; red hatched: transport properties; yellow: reaction parameters; green: apparatus related parameters)

Conclusion

To improve energy processes by finding better working substances it is important to identify the properties most relevant. However, the sensitivity of process performance on the properties alone does not allow deciding if the potential for improving the process is large. To do so it is also important to consider the range for the values the respective property realistically can assume. A previously presented[4] estimation method for the potentials related to different parameters has been extended in this work to also describe effects of transport properties. Furthermore, it has been demonstrated that the correlations between most properties are negligible and therefore independent variation of the parameters as used in the approach is reasonable.

Acknowledgments

The authors wish to thank Prof. Dr. Wolfgang Arlt for valuable discussions and his support of this work.

This project has been done within the framework of the *Energie Campus Nürnberg*.

References

1. Lampe, M.; Stavrou, M.; Schilling, J.; Sauer, E.; Gross, J.; Bardow, A., Computer-aided molecular design in the continuous-molecular targeting framework using group-contribution PC-SAFT. *Computers & Chemical Engineering* **2015,** 81, (Supplement C), 278-287.
2. Völkl, J.; Müller, K.; Mokrushina, L.; Arlt, W., A Priori Property Estimation of Physical and Reactive CO_2 Absorbents. *Chemical Engineering & Technology* **2012,** 35, (3), 579-583.
3. Preißinger, M.; Brüggemann, D., Auf COSMOtherm und PubChem aufbauendes High-Troughput-Screening potenzieller Arbeitsfluide für thermodynamische Kreisprozesse. Oral Presentation at the *Thermodynamik-Kolloquium* Bochum, **2015**.
4. Adametz, P.; Baumgärtner, A.; Kohler, T.; Obermeier, J.; Siebert, D.; Müller, K., Analysis of the Potential for Improvement of Chemical-Energy Transformation Processes. *Chemical Engineering & Technology* **2017,** 40, (6), 1115-1123.

494311 Life Cycle Optimization of Topside Process Design for Offshore Platform

Abstract:

During a life cycle of gas or oil well exploration, components of fluids and operating condition of the wells constantly changes. For this reason, the operating condition and control strategy of the topside process at either FPSO (Floating Production Storage and Offloading) or fixed platform needs to be optimized in real-time. In this study, a number of cases that might happen during the life cycle of gas well were considered for the design of top side process. In particular, dynamic simulation was adopted to consider equipment sizing and control logics. As a result, we were able to reduce unnecessary margin for the design of topside process and increase the efficiency of the process operation.

This research was funded by Ministry of Trade â€œDevelopment of integrated Interactive Model for Subsea and Topside System to Evaluate the Process Design of Offshore Platformâ€ (project No. 10060099)

Accounting for anisotropy of the packing to improve CFD simulation of liquid maldistribution in structured packing in an inclined column and validation

P. Béard, M. Fourati, P. Alix, IFP Energies nouvelles, Solaize, France
T. Maubert, C. Weiss, TOTAL, X. Courtial, G. Perdu, Prosernat, France

As natural gas always contains contaminants such as acid gases, in particular CO_2 and H_2S, acid gas removal is required to meet LNG (Liquefied Natural Gas) specifications. This process is performed in packed liquid-gas contacting columns operating under counter-current liquid-gas flow. In offshore conditions, the performance may be altered because of ship movement resulting in liquid maldistribution within packed beds.
In order to improve the design of such columns, this work investigates the liquid distribution in a pilot-scale column filled with a structured packing for different operating conditions. The column tilt angle (θ), diameter (D) and packed bed height (H) as well as the liquid mass flow rate (Q_L) were varied.
Unsteady 3D CFD (Computational Fluid Dynamics) simulations were carried out using ANSYS® FLUENT® with a two-phase liquid-gas Eulerian approach. The structured packing was modeled as a porous medium wherein gas-liquid momentum exchanges and liquid dispersion were taken into account. The previously developed CFD model dedicated to the studied packing [1] allowed to reproduce properly the sensitivity to the operating conditions [2]. Nevertheless, the liquid maldistribution was overestimated when the height of the column is increased. Thus, the model was improved to account for the packing geometry by blocking the flow in the direction perpendicular to the walls (Figure 1) leading to an anisotropic dispersion of the liquid phase. This led to a modified distribution of the liquid in the column (Figure 2) compared to the results obtained with the isotropic model.

Experimental measurements were carried out in a test rig installed on a hexapod moving platform allowing ship motion simulation (Figure 3) at Heriot Watt (HW) University. These data are used to validate the liquid maldistribution predicted by the numerical simulations.

Comparisons exhibit an enhanced agreement (Figure 4) with the more realistic description of the packing geometry developed in this study compared to previous results [2]. The predictivity of the new anisotropic CFD model remains very good for high values of the column tilt angle and the liquid mass flow rate (Figure 5) or if the column diameter is increased (Figure 6).

References

[1] Fourati et al., "Liquid dispersion in packed columns: Experiments and numerical modeling", Chem. Eng. Sci. 100, 2013

[2] Béard et al., "Liquid maldistribution in structured packing in an inclined column: CFD simulation and validation", 9th Int. Conf. on Multiphase Flow, 2016

Authors

Philippe Béard, IFP Energies nouvelles, philippe.beard@ifpen.fr

Manel Fourati, IFP Energies nouvelles, manel.fourati@ifpen.fr

Pascal Alix, IFP Energies nouvelles, pascal.alix@ifpen.fr

Thomas Maubert, TOTAL, thomas.maubert@total.com

Claire Weiss, TOTAL, claire.weiss@total.com

Xavier Courtial, Prosernat, xcourtial@prosernat.com

Gauthier Perdu, Prosernat, gperdu@prosernat.com

Figures :

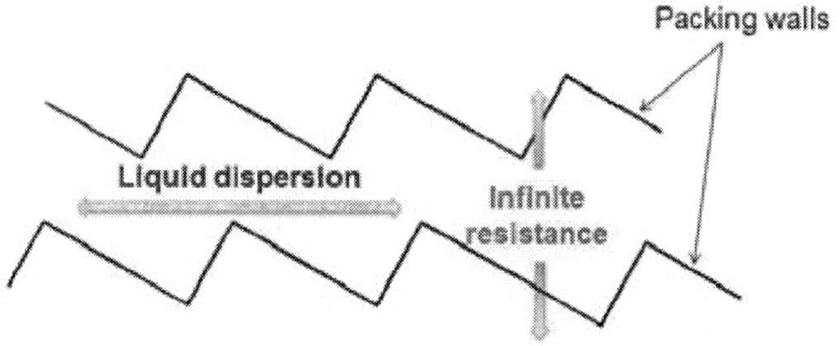

Figure 1 : Principle of anisotropic flow within the packing.

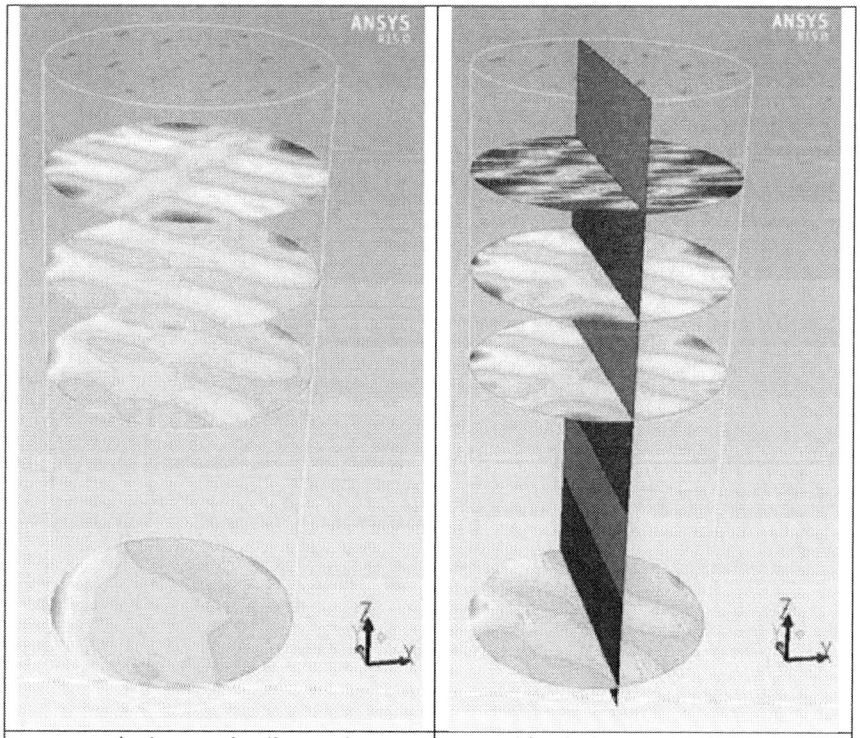

a) Isotropic dispersion	b) Anisotropic dispersion

Figure 2 : Comparison of the liquid volume fraction distributions between the isotropic and the anisotropic CFD models for D = 0.6 m, H = 1.2 m, θ = 3°, Q_L.

Figure 3 : Experimental test rig allowing ship motion simulation.

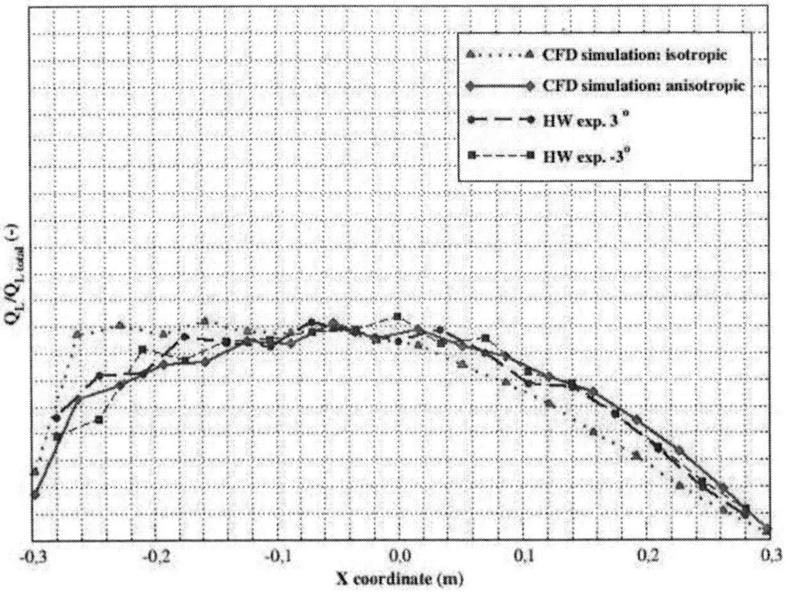

Figure 4 : Comparison between the isotropic and the anisotropic CFD models and validation for D = 0.6 m, H = 1.2 m, θ = 3°, Q_L

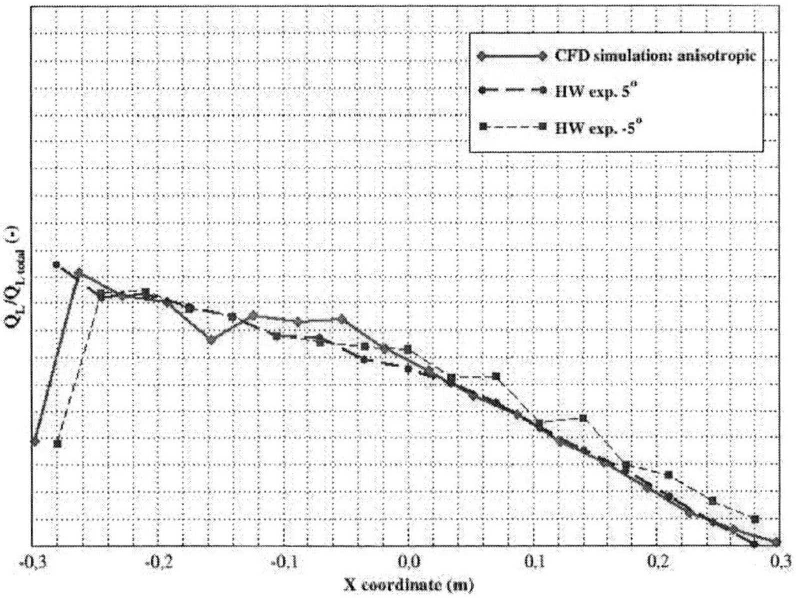

Figure 5 : Validation of the anisotropic CFD model for D = 0.6 m, H = 2.4 m, θ = 5°, 2 Q_L

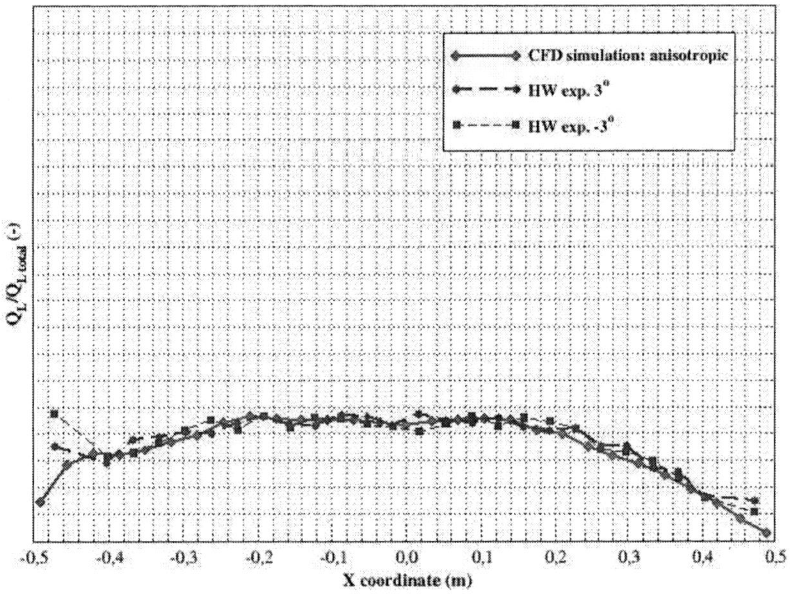

Figure 6 : Validation of the anisotropic CFD model for D = 1 m, H = 1 m, θ = 3°, Q_L

497367 Intensification of Cryogenic Lithiation-Borylation Though Use of a High Heat Transfer Loop Reactor

Abstract:

Lithium-hydrogen exchange is a common chemical transformation used to functionalize molecules, specifically to add boron to phenyl rings. One challenging class of these lithiations is the case where fluorine directs the lithium to the ortho position. These ortho-fluorine lithiated phenyl species can undergo highly exothermic decomposition through benzyne intermediates and are therefore run at deep cryogenic conditions. Not only does the intermediate need to be kept cold, but the subsequent borylation step also generates a substantial amount of heat that can promote the decomposition reaction. The challenge of handling the heat generation means operating this process requires good control of the temperature and rate of heat generation. In a batch process, this means slow addition of reagents at low temperatures. Changing operation from a batch to a continuous process presents the opportunity to increase the heat transfer rate and potentially support higher temperatures and faster mixing.

A small scale cryogenic flow chemistry platform was used to initially screen this reaction in a series of two tubular reactors.(1) At this milliliter scale, the lithiation was operated at -50 °C with the lithiated intermediate directly mixed with the boron source without degradation in a simple y-mixer. The reaction was further scaled up by approximately a factor of 3 to generate gram quantities of material and faced several challenges in the mixing section with respect to heat removal.(2) These were overcome by reconfiguring the mixing zones and better controlling heat removal from the reactor. These early results indicate that further scaling up of the process using the same configuration would present insurmountable problems.

We will discuss an additional 10x scale up capable of producing kilogram quantities of material over the course of a day. This required redesigning the borylation reactor to use a loop reactor with a heat exchanger inline to remove the heat of reaction rapidly. We will also show how careful design of the lithiation reactor would produce plug-flow performance even under laminar flow at the larger scale. The envisioned production scale continuous plant would have a 100x smaller footprint and reduction in potential chemical energy contained in the unstable intermediate. This would be safer and more economical than the equivalent batch process.

References:

1. Newby, J. et al. Design and Application of a Low-Temperature Continuous Flow Chemistry Platform. Org. Process Res. Dev. 2014, 18, 1211-1220.

2. Newby, J. et al. Reconfiguration of a Continuous Flow Platform for Extended Operation: Application to a Cryogenic Fluorine-Directed ortho-Lithiation Reaction. Org. Process Res. Dev. 2014, 18, 1221-1228.

497922 Design of Novel Microalgal Photbioreactor Using Computational Fluid Dynamics

Abstract:

The incessant rise of global CO_2 levels in the environment, depleting fossil fuel reserves and increased human & industrial activities are putting our ecosystem at risk. The search of renewable and sustainable options for generating fuels are on the rise. Microalgae is regarded as a green alternative owing to its several beneficial characteristics such as faster growth rates, CO_2 fixation abilities, etc. Moreover, microalgae has a tremendous potential for efficient CO_2 capture from point sources, biomass production for bio-product generation (e.g. biofuels, cosmetics, fodder and food etc.) and waste-water mitigation. However, one of the major bottlenecks in the microalgae-to-biofuel value chain lies in the growth of microalgae for commercial scale. Microalgae is cultivated in open as well as closed systems. However, the improved performance of a microalgal growth system is generally exhibited in closed photobioreactor which is governed by several physico-chemical factors such as light, pH, mixing, aeration, CO_2 etc. It is generally true that hydrodynamics plays a significant role by improving the mixing of nutrients, gases (improved mass transfer), avoiding the creation of dead zones, improved light penetrability and reduced sheer stress on the biological cells as well as reduced power consumption. In field experiments as well as computer simulation studies, the most common types of PBRs which are thoroughly investigated are bubble column and air-lift reactors. However, there is a scope to enhance flow patterns in PBRs using advanced CFD techniques. Computational Fluid Dynamics (CFD) enables determination of the hydrodynamics of the culture medium and mixing efficiency according to various PBR designs which is otherwise difficult to measure in field experiments. In this work Eulerâ€"Lagrangian model has been used to evaluate the turbulence mixing for two phase flow in airlift PBR using commercial CFD software. Three reactor configurations have been proposed with variation in draft tube length and type for assessing its effect on hydrodynamic characteristics. Parameters such as flow rate (Q), mean velocity (**u**), and mean Turbulent Kinetic Energy (TKE) were characterized for each configuration. The dimensions of the novel design were optimized using iterative design procedure. In all experiments, tap water was used as the liquid phase and was operated in a batch mode, while compressed air (CO_2 rich) was used as the gas phase and was operated in continuous mode. The hydrodynamics properties such as axial dispersion coefficient, mixing time and circulation time were calculated. The experimental results and CFD simulations were thoroughly compared for the novel PBR with that of conventional airlift reactor. In addition to performance evaluation of the novel PBR, a green microalga *Chlorella* sp. is cultivated in the PBR using air/flue gas and its growth characteristics were determined. The performance of the novel PBR was compared with that of the conventional airlift PBR. The results showed that the novel design had superior liquid circulation properties that resulted in a regular and alternative exposure to light and dark regions of the PBR. Also, CFD led to improved CO_2 mass transfer efficiency, biomass production, nutrient and light utilization profile.

501874 Process Intensified and Direct Production of Gasoline from Syngas

Abstract:

An Abstract Submission to 2017 AICHE Meeting

Session: Process Intensification By Enhanced Heat and Mass Transfer

Topical Area: Process Intensification

Process Intensified and Direct Production of Gasoline from Syngas

Xinquan Cheng[a], William R. Yantz Jr.[a], and Bruce J. Tatarchuk[a, b,*]

[a] Center *for Microfibrous Materials Manufacturing, Department of Chemical Engineering, 212 Ross Hall, Auburn University, AL 36849*

[b] *IntraMicron Inc., 368 Industry Dr., Auburn, AL 36832, USA*

* *Corresponding author. Tel.: +1 334 844 2023; fax: +1 334 844 2065; E-mail address: Tatarbj@auburn.edu*

Abstract

The increasing demand for liquid fuels draw people's attentions to the synthetic fuels. Currently, there are mainly two approaches to synthesize liquid fuels from syngas. One is Fischer-Tropsch synthesis (FTS) and another one is syngas-to-methanol first and followed up with methanol-to-gasoline (MTG). Those two approaches are well-known techniques for converting syngas to liquid fuels. However, FTS process produces a broad spectrum of products, which requires complex and expensive subsequent processes to upgrade the initial products in order to meet liquid fuel specifications. The MTG approach is a multiple step process. Typically, in the first reactor syngas is converted to methanol, and followed with methanol dehydration reactor to produce the intermediate product dimethoxyethane (DME), and then DME successively converted to hydrocarbons in the reactor packed with zeolite. It must be noticed that all those reactions are highly exothermic reactions, which producing extremely amount of heat during the reaction processes. Therefore, the reaction thermal management is always a big issue for syngas to liquid fuels.

Here, we present a single reactor system to directly convert syngas-to-gasoline (STG) by using a novel catalyst structure called Microfibrous Entrapped Catalyst (MFEC) structure. The MFEC structure was developed by our group, which is a catalyst network supported by micron-sized metal fibers with high thermal conductivity. By supporting small catalyst particles into a high voidage matrix of microfibers the catalyst is suitable for fixed bed applications without severe pressure drop or bed channeling encountered in small particle packed beds. Previous efforts on this unique catalyst structure have shown significant enhancement in intra-bed heat transfer and mass transfer.[1-4] Basically, we packed a large tubular reactor (34.0 mm I.D.) in three zones by using Cu MFEC. As we can see from Figure 1, the first zone (Zone-1) is a mixed catalyst zone for the production of DME from syngas. The second zone (Zone-2) is a relative inert zone mainly used to separate Zone-1 from Zone-3 since those two zones have different temperature. In order to convert all the unreacted methanol from Zone-1, we packed γ-Al_2O_3 in Zone-2 as a methanol dehydration catalyst. The third zone (Zone-3) is a zeolite zone packed with ZSM-5 to convert the DME to gasoline. This kind of multi-zone reactor design can separate the zeolite from the methanol synthesis catalyst, because the proximity of those two catalyst components is crucial factor for the product selectivity of STG process.[5] If those two components are close to each other, the lower olefins, which are the essential intermediates from methanol to aromatics, diffuse from the zeolite to the methanol synthesis catalyst, where they are easily hydrogenated to lower paraffins. It turns out that most of the products are short chain paraffins.[6]

However, with this simple multi-zone reactor design, the olefinic intermediates are preserved for subsequently oligomerization reactions. Under a typical reaction conditions (P=30 atm, GHSV=1000 h^{-1}, H_2:CO=1:1), the $C_5 - C_{10}$ hydrocarbon selectivity can reach 39.0 wt% with a relative high CO conversion (73.2 %). In addition, the Cu MFEC structure demonstrated a very low radial temperature gradient (around 5 °C) for a 34.0 mm I.D. reactor and insignificant amount of pressure drop. While with the same test conditions, the comparative packed bed reached a very high radial temperature gradient (around 140 °C) and significant amount of pressure drop. Considering these attributes and advantages, the Cu MFEC multi-zone reactor system is a very efficient process intensification technique for the direct production of gasoline from syngas.

Figure 1. The STG reactor design (all the catalysts particles entrapped in Cu MFEC in Zone-1, Zone-2, and Zone-3 are in 60 – 80 mesh)

References:

[1] M. Sheng, H.Y. Yang, D.R. Cahela, B.J. Tatarchuk, J. Catal. 281 (2011) 254-262.

[2] M. Sheng, H.Y. Yang, D.R. Cahela, W.R. Yantz Jr., C.F. Gonzalez, B.J. Tatarchuk, Appl. Catal. A-Gen. 445-446 (2012) 143-152.

[3] M. Sheng, D.R. Cahela, H.Y. Yang, C.F. Gonzalez, W.R. Yantz Jr., D.K. Harris, B.J. Tatarchuk, Int. J. Heat Mass Tran. 56 (2012) 10-19.

[4] X.Q. Cheng, H.Y. Yang, B.J. Tatarchuk, Catal. Today 273 (2016) 62-71.

[5] S. Sartipi, M. Makkee, F. Kapteijn, J. Gascon, Catal. Sci. Technol. 4 (2014) 893-907.

[6] K. Fujimoto, H. Saima, H. Tominaga, J. Catal. 94, (1985) 16-23.

487118 Modelling of Time-Dependent Interfacial Properties Due to Chemical Equilibrium Reactions in Demixed Fluid Systems

Abstract:

Modelling of time-dependent interfacial properties due to chemical equilibrium reactions in demixed fluid systems

Andreas Danzer, *Sabine Enders*

Karlsruhe Institute of Technology, Institute of Technical Thermodynamics and Refrigeration Engineering, Engler-Bunte-Ring 21, D-76131 Karlsruhe, Germany.

Email: andreas.danzer@kit.edu; sabine.enders@kit.edu

Key words: reactive extraction, impact of chemical reactions on interfacial properties, model calculations

Abstract:

Reactive extraction, the integration of reaction and extraction within one unit operation, is a promising process in a variety of applications. Compared with the conventional process of carrying out reaction and separation sequentially, the in-situ removal of products formed in the reaction zone can favorably influence conversion and selectivity, especially of equilibrium reactions. Together with a significantly reduced effort in the task of separation, this can result in major energy as well as capital cost savings. However, there are feasibility constraints resulting both from thermodynamics and chemistry. In addition, the process intensification leads to higher complexity and thus the need for research in terms of modelling productiveness. Because of that model calculations of hypothetically ternary liquid-liquid reactive extraction systems were performed in this work. The liquid-liquid phase equilibrium was calculated with the well-known, quite simple Margules activity model, the interfacial properties were modelled with the Margules equation in combination with the Density Gradient Theory [1-4] (DGT). It was assumed that the chemical reaction occurs in a A B manner, whereas the solvent C does not participate in the chemical equilibrium reaction. Furthermore, the assumption was made that mass transfer occurs at speed of infinity, so the chemical reaction is rate determining, in contrast to the reach of phase equilibrium. The empirical constants of the Margules equation $A_{i,j}$, the so called influence parameter κ of the DGT as well as the chemical equilibrium constant $K=k_1/k_2$ were specified due to the need to serve as a model. For real systems, $A_{i,j}$ must be fitted to experimental equilibrium data of the binary subsystems, the influence parameter κ has to be fitted at one experimental data point of the interfacial tension of each binary system with a miscibility gap and K is determined by the chemical reaction. It was found that the time, to reach chemical equilibrium in addition to the phase equilibrium, is strongly dependent on the shape of the miscibility gap as well as on the slope of the tie lines. Furthermore, the interfacial tension of the demixed system decreases over time, because of the fact that the product of the chemical equilibrium reaction act as a solubilizer. The slope of the interfacial tension over time strongly corresponds to the quotient of the reaction rate of forward and backward reaction. The new theoretical framework allows to the first time the calculation of time-dependent interfacial tensions and can further transferred to real systems.

Literature:

[1] J.W. Cahn, J.E. Hilliard, J. Chem. Phys. 28 (1958) 258.

[2] C.I. Poser, I.C. Sanchez, Macromolecules 14 (1981) 361.

[3] T. Grunert, H. Rudolph, S. Enders, Zeitschrift für Physikalische Chemie 227 (2013) 1.

[4] T. Grunert, S. Enders, Fluid Phase Equilibria 381 (2014) 46.

495073 Impact of Internals on Bubble Column Performance

Abstract:

Impact of internals in a bubble column reactor is a widely studied topic especially for the interesting hydrodynamics involved. Though numerous experiments have been performed utilizing horizontal, vertical internals and other packing materials, their direct, individual impacts on the bubble column performance is complex and not fully understood. Utilizing effective CFD simulations to recreate the flow conditions has become a necessity to understand these dynamics better. Based on the reaction, operating regime and flow conditions, it is possible to drive the back mixing, liquid recirculation and other parameters favorable to our requirements, by modifying the internal structures. We made an attempt to quantify the performance of various internal designs (taken from literature), for a given column, based on selected hydrodynamic parameters. With a better understanding of the flow patterns and their response to a number of internal structures, we take this forward and propose new designs which are tailored specifically for the reaction conditions and maintaining the flow in a particular regime (bubbly flow regime in our case) even at higher U_g thereby successfully delaying transition. We validate them by 3-d printing these designs and testing out their performance on a laboratory scale bubble column.

500888 Prediction of Tablet Dissolution By Process Parameters in Continuous Manufacturing

Abstract:

Tablet dissolution is an attribute that is critical to the product quality and should be well controlled in order to produce an acceptable pharmaceutical product. Tablet dissolution testing is a lengthy procedure that is time-consuming, expensive, and requires media preparation. In addition, it is a destructive test, and the sample will no longer be available for further characterization. Hence, there has been a growing interest in the development of non-destructive predictive tools for tablet dissolution to enable the real time release testing (RTRt) strategy. Application of such a strategy would not only lower laboratory and personnel costs but also increase product quality. A statistical model for predicting tablet dissolution profiles will enable rapid techniques in continuous assessment of tablets ensuring the desired quality attributes.

In this work, we developed a statistical model to predict tablet dissolution profile in a continuous direct compaction manufacturing line. Tablets consisted of model API, Acetaminophen (APAP), lactose monohydrate as the filler, and magnesium stearate as lubricant. Multivariate effects were analyzed using Quality-by-Design methods. The experiment was based on a fractional factorial design with 30 experiments, including 3 repeated center points, to study the influence of four process variables – drug concentration (%), compaction force (KN), blending speed (rpm), and feed frame speed (rpm) – on the final product. Dissolution profiles of 180 tablets (six tablets for each experimental run) were obtained and fit into the Weibull model (including parameters of α and β), and the values of α and β were calculated. Multi variate analysis of variance (MANOVA) was used to determine the statistical differences between the dissolution profiles. The results indicated that compaction force and API concentration were the main parameters affecting tablet dissolution in this process. Therefore, a predictive model was developed to correlate the dissolution model parameters α and β to the process parameters. The model was verified using an independently manufactured validation set. The values of α and β of the validation set were estimated by the model. The dissolution profiles of these tablets were predicted using the estimated values of α and β. The predicted dissolution profiles obtained from the model were in agreement with the experimental results using the similarity factor f_2 value greater than 50. This work establishes an approach to the real time release testing (RTRt) of pharmaceutical tablets and the possibility of assessing the quality characteristics of tablets non-destructively.

501282 Integrating Sensors for Monitoring Blend Content in a Pharmaceutical Continuous Manufacturing Plant

Abstract:

In a pharmaceutical manufacturing process, Critical Quality Attributes (CQAs), as defined by the FDA, need to be monitored not only for the final product but also for intermediates. Blend uniformity of powders is one such attribute that needs to be measured to ensure the quality of the final product. In-line sensors, to monitor the blend content of the powders, were implemented within a Direct Compaction (DC) continuous tablet manufacturing line. In most cases, since the primary ingredient of interest is the active pharmaceutical ingredient (API), the amount of the API was monitored/predicted over the course of manufacturing. For the calibration model building process, a unique setup involving dynamic powder spectral acquisition method was used. This setup was aimed at mimicking the powder flow characteristics within the manufacturing line. A Raman probe and a portable NIR were used as sensors and they were placed at the exit of the blending process before the tableting stage. The performance of the two sensors and their respective models were evaluated in terms of accuracy, precision, operating range, measurement frequency, placement, reliability, robustness. Additionally, their performances were validated by off-line traditional analytical measurements.

501455 Scientific and Regulatory Considerations for Developing the Control Strategy of a Continuous Manufacturing Process

Abstract:

Advances in manufacturing technology over the last decade have provided the pharmaceutical industry opportunities for adopting innovative ways to develop flexible, efficient and robust processes. In recent years, continuous manufacturing has emerged as a new technology for the manufacture of pharmaceuticals. Compared to traditional batch process, continuous manufacturing has various advantages, such as, shorter processing times, smaller facility footprint, etc. On the other hand, it has unique elements that lead to uncertainty on how to implement a continuous process within the current regulatory framework. The Food and Drug Administration supports the adoption of innovative technology and is communicating with pharmaceutical companies and academia to facilitate the development of continuous processes. This presentation will discuss several scientific, current good manufacturing practice (CGMP) and regulatory considerations for developing the control strategy of a continuous manufacturing process.

500629 Uncertainty Quantification and Global Sensitivity Analysis of Drug Delivery Primary Containers to Understand Process Capability and Key Risk Factors

Abstract:

A simple high-fidelity model was created to predict the plunger position and air gap in a pre-filled syringe primary container to be used as a component in a combination product. The product-container system has six sources of uncertainty and two uncertain variables related to operating conditions. An uncertainty analysis was carried out to compute the joint distribution of plunger position and air gap performance parameters with the goal to understand the operating ranges for these CQA (critical quality attributes). In order to relate these to operational practices, process capability indices (Cpk) were calculated via a naïve approach where the marginal plunger-depth distribution was fit to a normal distribution in order to enable the index (Cpk) calculations. This simplified approach was compared with a full propagation of factor uncertainties via the rigorous nonlinear model, which resulted in a remarkable difference in the CpK values. This indicates that the Gaussian approximation was unsatisfactory and more refined techniques are required. As a final step, a global sensitivity analysis based on the Sobolâ€™ indices technique was carried out in order to identify the key factors affecting process capability and inform continuous process improvement initiatives. Overall, four of the eight factors were found to be unessential (non-key).

This work demonstrates the use of rigorous mathematical process models and advanced uncertainty/sensitivity techniques to shed light on design and operation performance characteristics (QbD and QbC). The model and techniques were implemented in a web-based deployment platform in order to leverage the rigorous models/techniques and share these simple-to-use applications with all users across the organization, regardless of their background.

500701 Nucleation Behavior of Eszopiclone - Butyl Acetate Solutions from Metastable Zone Widths

Abstract:

Experimental data on the critical supercooling ΔT_{max}, a measure of metastable zone width (MSZWs), as a function of saturated temperature T_0 and cooling rate R were analyzed to describe the nucleation behavior of unseeded eszopiclone-butylacetate solutions. The Nývlt's semiempirical model and Sangwal's theory were employed, respectively. And moreover, we propose a modified model based on the Sangwal's theory in which the nucleation parameters are not dependent on saturation temperature T_0 and nucleation temperature T_1 which can fit the experimental data better than Sangwal's theory. Furthermore, the pre-exponential A and solid – liquid interfacial energy γ can be obtained by employing this modified model, thus, sheds light on the relationship between nucleation parameters and the cooling rate R. Meanwhile, based on the Classical Nucleation Theory, with the estimated interfacial energies from the slope in the modified model, the critical nucleus size and critical Gibbs free energy were calculated and discussed. Then based on the Sangwal's theory, the nucleation activation barrier is found to be dependent on ln R with an exponential relationship, hence, the relationship between nucleation barrier and pre-exponential A can be constructed. At last, the effects of agitation rate and working volume on MSZWs were investigated and discussed.

Development, Implementation, and Use of an On-line Laser Diffraction Particle Size Method in a Spray Drying Manufacturing Process

Pharmaceutical Discovery, Development and Manufacturing Forum
2017 AIChE Annual Meeting

Martin Warman, Martin Warman Consultancy Ltd, and Alon Vaisman, Malvern Instruments

Malvern Panalytical

Why is this important - background

- The Biopharmaceutics Classification System (BCS)
  - Class I - High Permeability, High Solubility
  - Class II - High Permeability, Low Solubility
  - Class III - Low Permeability, High Solubility
  - Class IV - Low Permeability, Low Solubility
- In general, product performance for Class I + III compounds is driven by API concentration and tablet disintegration
- However product performance for Class II + IV compounds is driven by many more factors

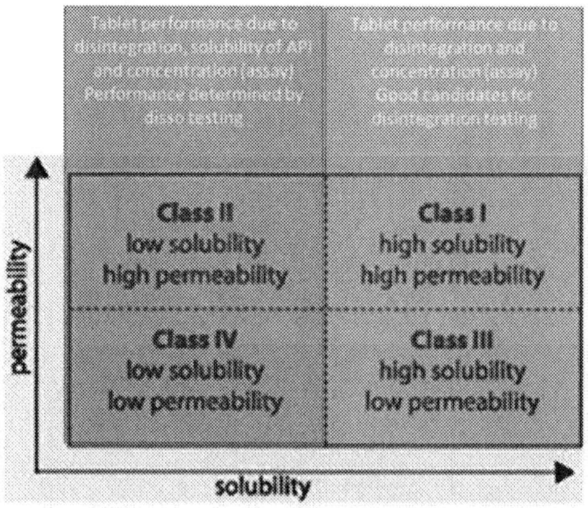

Malvern Panalytical

What is disintegration and weight are not enough

- For BCS II and IV API can't use simple indicators of tablet performance
  - Need to establish 'rate of release' data
  - Disso testing
- However the disso performance is a combination of numerous chemical and physical properties of the matrix
  - Concentration
  - Solubility
  - Hardness
  - Density
  - Porosity
  - Particle size
  - Inter-particulate forces
  - Etc
- Many of these are impacted by the physical API properties but also the intermediate particle properties
  - Spray Dried Dispersion
    - % crystallinity
  - Granulation

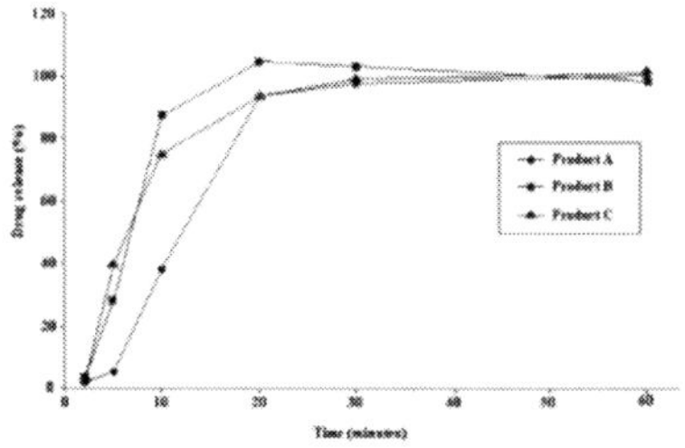

Malvern Panalytical

Case Study - High-level description of Process

- Crystalline API production
  - Reaction steps, crystallization, drying
- Amorphous API spray dried dispersion
  - Spray solution prep, spray drying, secondary drying
- Formulation
  - Pre lube blend, post lube blend
- Compression
  - Compression, film coat
- Described as full QbD, with elements of real-time release

Malvern Panalytical

Identifying steps critical to tablet performance

- For this product we could show the product performance was a result of
  - Tablet compression
  - SDD bulk density and particle size
  - SDD particle size could be controlled byt the spary drying process

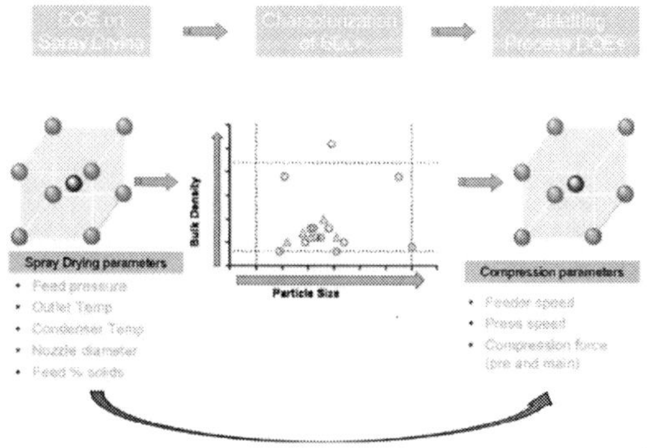

Malvern Panalytical

Identifying control models

- Regression analysis showed
  - The particle size of the wet SDD was determined by nozzle diameter and feed pressure
  - Whereas bulk density related to speed of drying, with droplets that dried fast having symmetrical shape (often with entrained gas bubbles) and droplets that dried slowly having high surface area/higher bulk density
- The key message is these could be controlled independently

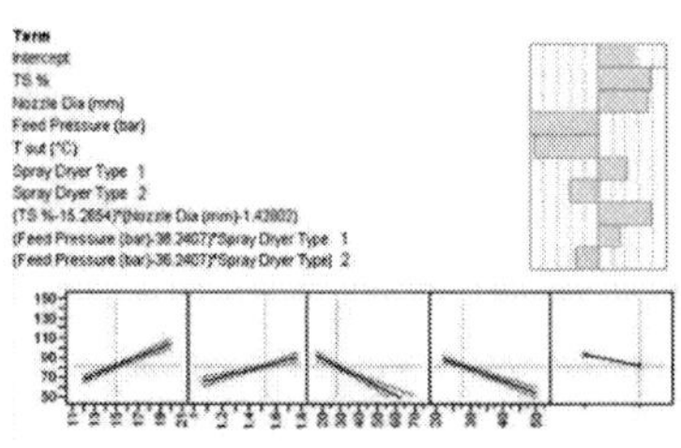

Particle size

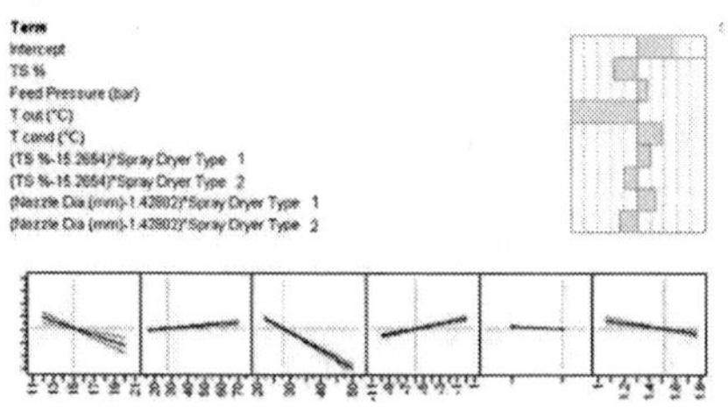

Bulk density

Malvern Panalytical

Graphical Illustration of Design Space

- More so we could demonstrate a three dimensional Design Space for product performance (disso)
- Control strategy complexity pyramid
  - First published by FDA ETT in the AAPS Journal, Vol 16, Number 4, July 2014
  - Included in the paper, Modernizing Pharmaceutical Manufacturing: From Batch to Continuous Production, Journal of Pharmaceutical Innovation, Vol 10, Issue 3, Sept 2015,
- The product filling pre-dates the complexity pyramid by 4 years but in hindsight the product was filled with a level 2 CS, but with the process knowledge to allow level 1 control

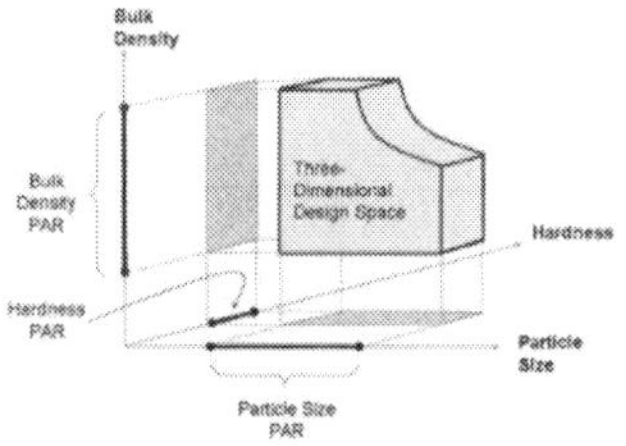

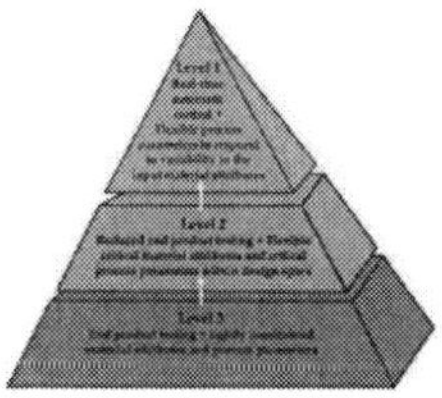

Malvern Panalytical

Control Strategy Implementation

- Dashboard sits on plant control system showing
- Predicted disso ranges shown to SDD process operator
  - If particle size and bulk density stay in the green range, the tablet compression set points are wide
  - If particle size and bulk density are in the yellow range, the tablet compression set point becomes critical
  - If the particle size goes red, there are no approved compression set points that will give acceptable product performance
- In reality the process runs in a NOR where we are always making good product
- Key message is we have real-time indicator of CQA predictions from PAT and parametric data

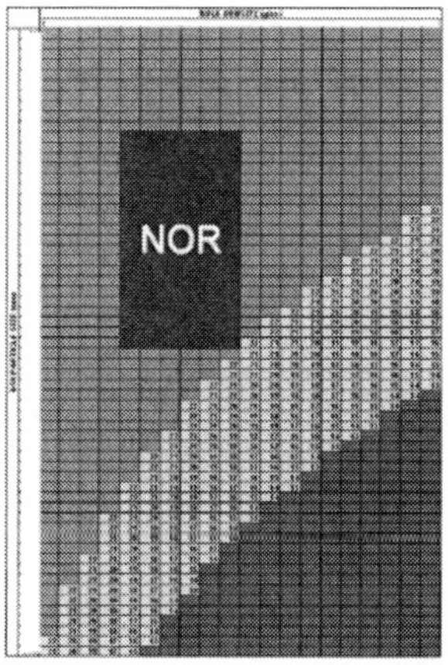

Malvern Panalytical

Key Question
How do we get that data?

Malvern Panalytical

Particle Size Application
Ensuring Representative Measurement

- Appropriate selection of sampling location
- Powder flow is linear (post vortex breaker)
  - However this means it is the 'shadow' of the vortex breaker
- Tried all commercial sampling systems
  - J-type, single funnel, etc
- Poor correlation with off-line data
- Implemented sampling arrangement is the result of empirical testing of various restrictions and funnel configurations
- Has been proven to generate the equivalent results
- Covers 25% of the cross sectional area, i.e. 25% of production runs through the analyser!

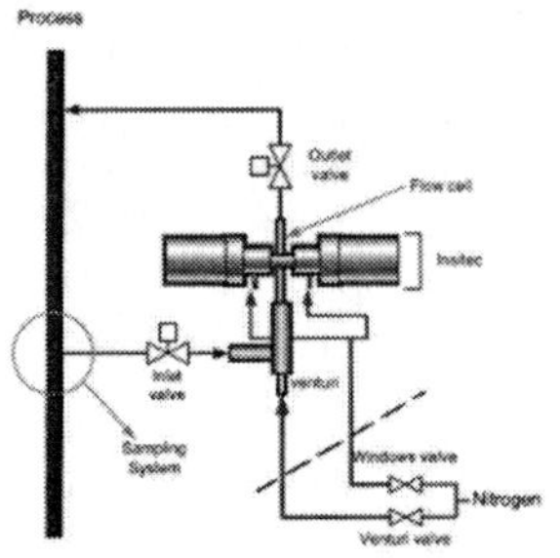

Malvern Panalytical

Prediction models

- First thing to realize is these the regression analysis is to SDD particles which had already been through the secondary dryer
  - Output of spray drying still has some residual solvent
  - Need to predict 'dry' values from 'wet' data
  - Models using process data predict dry particle size and dry bulk density from achieved set-points
- Secondly off-line methods are not the same as PAT methods
  - PAT methods need to predict off-line values as off-line values have been used in the control models
- In terms of particle size
  - Off-line method uses same fundamental measurement technique (laser, scattering, Mastersizer)
    - BUT Mastersizer uses the Fraunhofer approximation, and the Insitec used Mie scattering theory
    - And the off-line method uses a pressure titration to distribute the particles, whereas the PAT method uses flow titration (velocity)
  - PAT method needs to predict off-line values

Malvern Panalytical

Initial correlation between PAT and off-line

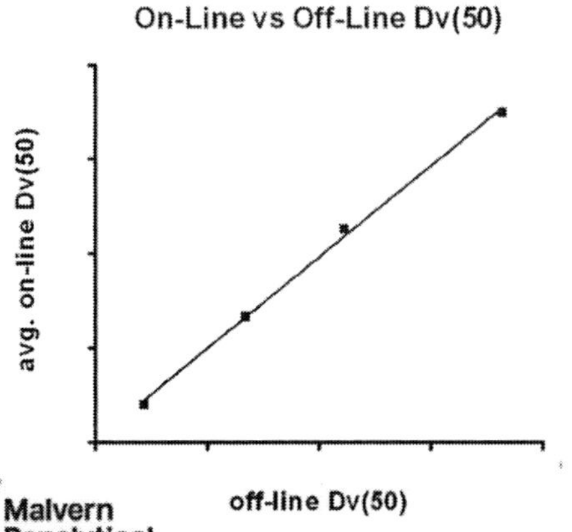

	Parameter	Value
Hardware	Online Laser Diffraction System	Insitec X
	Lens	200 mm (1-400 micron focal length)
	Venturi	90 degree
	Flow rate	9.5 Nm^3/hr
Material Settings	Model	Mie Scattering
	Particulate Refractive Index	1.5 + 0.1i
	Media Refractive Index	1
	Particle Density	1
Data Analysis	Min Particle Diameter (microns)	0.1
	Max Particle Diameter (microns)	600
	Scattering Threshold	8
	Multiple Scattering	Yes
	First Scattering Ring	5
System Suitability	Averaging	8-minute
	Transmission	≥ 80 %
	Counts at Sc(15)	≥ 500

Malvern Panalytical

How do we know this works?

- Example of a process change
- Correlation of process parameters to intermediate attribute means we can change process parameters to cause desired change in attribute
- Make a change expected to move the Dv50 by 3um
  - Dv50 responds immediately and stabilizes in 8mins (one block of data)
- More importantly we can set the press parameters to give the disso performance we require.

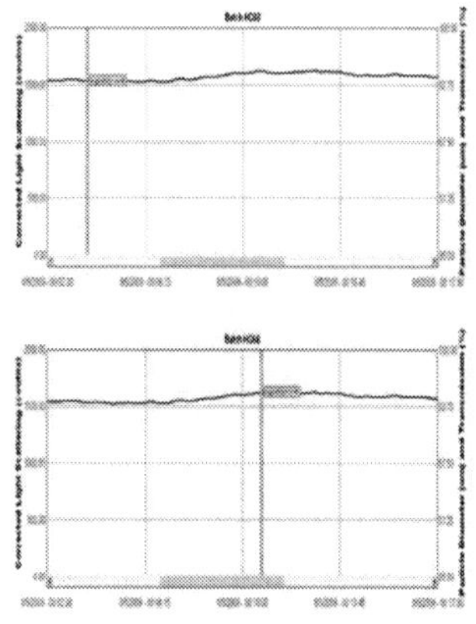

Malvern Panalytical

13

What does this all look like together?

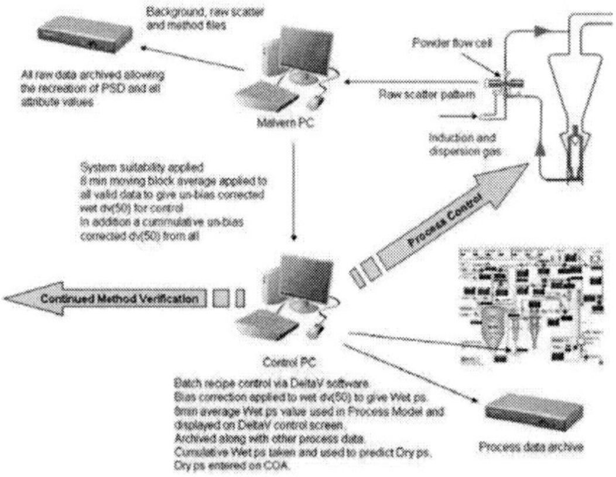

Malvern Panalytical

PAT Method Lifecycle

- Trend analysis of key performance indicators
  - Difference between predicted values (from achieved set-points)
    - 'parallel testing' of soft-sensor methods
    - Method updates
  - Difference between PAT measured values to off-line
  - Causes of PAT method system suitability (transmission at Sc15)
  - Predicting how process variance would impact the above
    - Proactive triggers for method updates

Malvern Panalytical

Differences between PAT and off-line methods

- Early PAT prediction methods of dry, off-line data where simple linear regressions (bias adjustments)
- However over time, and by using expanded ranges, we could show that the slope of all three PSD attributes (d(50), d(10) and d(90)) were not 45 degrees
- They are close enough to predict in the operational ranges
  - Fixed bias changed to slope adjustment
  - Fixed was OK, slope was better

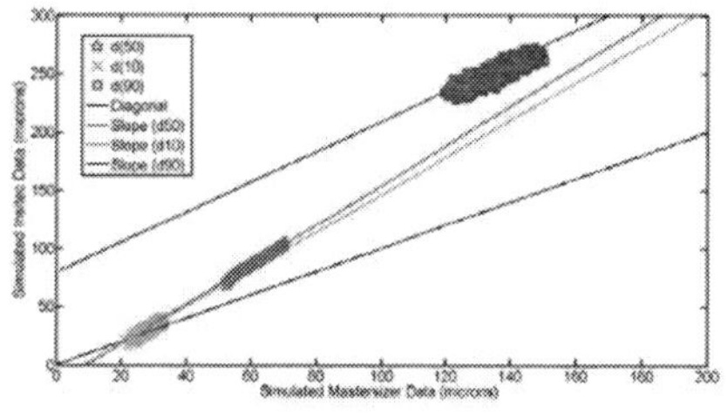

Malvern Panalytical

Characterizing Sc15 and transmission 'failures'

- PAT method uses transmission and counts on Sc15 as a system suitability
- Initially these were simple acceptance values
  - Found to be more sophisticated
- When transmission passes the PSD is mono-modal and SC15 has highest count
- As actually the system suitability should be on signal 'shape'

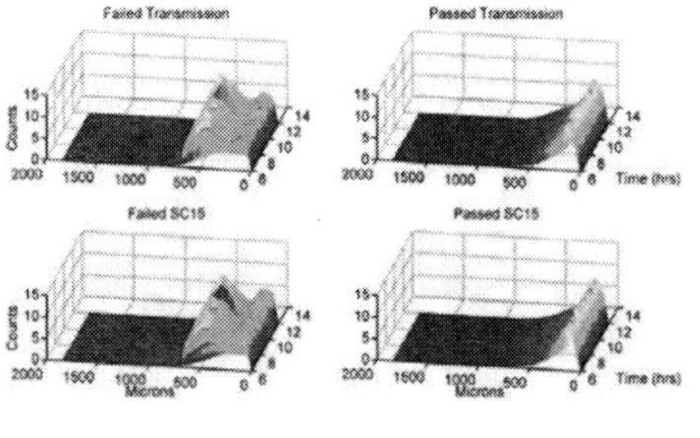

Malvern Panalytical

Impact of viscosity of d(50)

- During development an acceptance range for API/polymer solution was defined
- During production it was found that the viscosity of API/polymer solution varied (within range) and that if it varied the d(50) that would be generated changed
  - Suggested from regression analysis (positive correlation with %TS)
  - Fine if we stay in the narrow operating range, but if the range was expanded then a set-point correction should be applied to correct for cP
- However during the life cycle of the product various factors were identified which impacted viscosity
  - New acceptance tests for raw materials used in spray solution

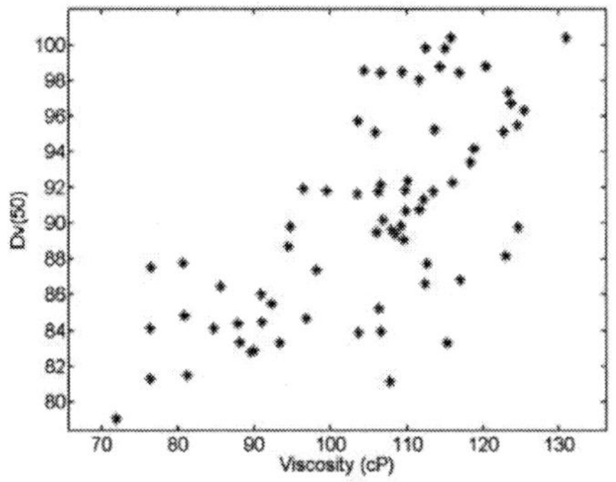

Malvern Panalytical

Continued application development

- Product underwent tech transfer to a bigger spray dryer, in a different facility
- Insitec installation moved upstream
  - Directly below drying tower
  - Before cyclone
- Measurement across entire discharge pipe
- Even then the installation required fine tuning
  - Lens were originally parallel, found to be reflection (multiple bounces) impacting scatter pattern
  - Original installation was across the center of the pipe, Insitec now rotated to look 'up-down' rather than 'left-right'
    - To account for stratified flow in the pipe
  - Method updated to include setting the 'zero' when process is spraying solvent
    - Changes refractive index

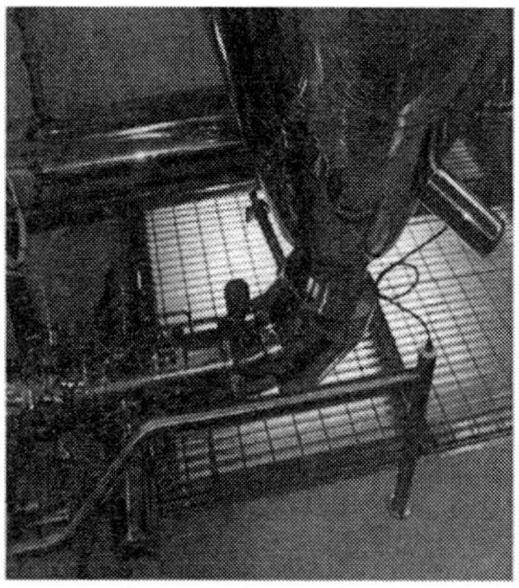

Malvern Panalytical

Conclusions

- In general, understand the inputs to tablet performance
  - Fishbone/Ishikawa diagrams are very useful
- Collect the data (either directly or via modelling) to have representative inputs into those diagrams
  - Use high frequency data collection rates to for control models and design space time limits, averaged data to determine product quality attributes
  - Justify data collection and decision frequency
- Make sure the data is weighted appropriately
  - For example in tablets with high drug load, weight variance is more important than low drug load, where variance in concentration has a bigger impact
- If you do, then full real-time release is achievable and has been achieved.

Malvern Panalytical

Acknowledgements

- Vertex
  - Joseph Medendorp, John Bric, Kelly Tolton, Greg Connelly
- Hovione
  - Filipe Gaspar, Nuno Matos, Filipe Neves
- Malvern Panalytical
  - Sensors and automation solutions team

Thank you for your attention.....

Malvern Panalytical

496216 Assessment of Economic Potential of Furfural Platform

Abstract:

Furfural is identified as a potential bio-based building block by the department of energy. A suit of more than 80 chemicals that can be directly and indirectly derived from furfural have been identified. It is a daunting task to evaluate the economic potential of all of these furfural transformations. We, therefore, qualitatively screened and selected following seven chemicals that have near-term commercial potential: tetrahydrofuran (THF), tetrahydrofurfuryl alcohol (THFA), furfuryl alcohol (FA), maleic anhydride (MA), 2-methyltetrahydrofuran (MTHF), levulinic acid (LA), and 1,5-pentanediol (PDO). The minimum selling price (MSP) of furfural produced via acid dehydration of C5 sugar and the seven selected potential furfural derivatives were determined. The MSP of furfural produced in a plant with a capacity of 2000 metric ton per day of bone dry wood was estimated at $0. 78 per kg. This estimation indicates that the plant size of platform chemical production and the MSP are comparable to the proposed feasible size of lignocellulosic based biofuel production plant and the market price of super commodity chemicals, respectively. The estimated MSP ($/kg) of THF, THFA, FA, MA, MTHF, LA, and PDO are 1.47, 1.1, 0.94, 1.3, 1.54, 1.47, 1.28, respectively. The comparison of calculated MSP of furfural derivatives against the market prices has shown that all analyzed production processes are economically viable except the conversion of furfural to levulinic acid. The comparison of economic and modeling analysis of analyzed production pathways of furfural derivatives enable us to determine general rules of thumb necessary to screen early stage transformations of potential bio-based building blocks to various secondary commodity chemicals.

496231 Economic and Modeling Analysis of Product and Energy Driven Biorefineries Based on Thermal Deoxygenation Pathway

Abstract:

Biorefineries can be classified into product- and energy-driven biorefineries. The thermal deoxygenation pathway invented by University of Maine researchers offers flexibility in use of feedstock, which allow us to construct both product- and energy-driven biorefineries. In the product-driven biorefinary, lignocellulosic feedstock such as wood is converted primarily to furfural (a potential bio-based building block) along with co-products biofuel and biomaterials char and activated carbon. In the energy-driven biorefinary, cellulose rich feedstock such as cardboard is mainly used to produce biofuel along with the co-product levulinic acid (a potential bio-based building block). In this study, we performed economic and modeling analysis of product- and energy-driven biorefineries based on thermal deoxygenation pathway to determine the best type of a biorefinary in terms of economic returns. Greenfield and brownfield plant site scenarios were considered in the economic assessment of product- and energy-driven biorefineries. ASPEN plus simulation software was used to model each type of biorefinary. Capital and operating costs were estimated for a plant capacity of 2000 metric ton per day of bone dry biomass feedstock, using the simulated material and energy balances. The discounted cash flow analysis with an internal rate of return of 10% was employed to estimate the minimum selling price of bio-based products produced in both product- and energy-driven biorefineries. Finally, economic viability of producing activated carbon from char rather than burning it to produce energy was assessed. Determining which type of biorefinary is the best in terms of economic performance will enable us to effectively utilize limited resources for the development of economically viable biorefineries that eventually accelerate the growth of a bio-based economy.

496935 An Algorithm for Integrated Design of Organic Rankine Cycles

Abstract:

An algorithm for integrated design of Organic Rankine Cycles

Uku Erik Tropp, David H. Bowskill, Smitha Gopinath, George Jackson, Amparo Galindo, Claire S. Adjiman

Department of Chemical Engineering, Centre for Process Systems Engineering, Imperial College London

Organic Rankine Cycles (ORCs) are able to utilize waste heat or low quality heat to generate power. Thus, the use of ORCS can lead towards sustainable and environmentally friendly energy systems. The optimal operating conditions of the ORC depend on the specifications of the process (for e.g., temperature of the heat source), constraints of the system (for e.g., amount of coolant available) and the choice of working fluid. Optimal cycles may be designed by simultaneously choosing the fluid and the operating conditions, subject to the constraints. Such an integrated design of materials (molecules) and processes is also referred to as Computer Aided Molecular and Process design (CAMPD) [1]. In the ORC CAMPD, process decisions are usually continuous, whereas material (molecular) decisions are inherently discrete. The ORC working fluid (a pure-component is considered here) may be represented by the number of occurrences in the molecule of each functional group (in a given set of groups) and the connectivity between these occurring groups. Given a property prediction method, that translates the molecular composition of the fluid to physico-chemical properties of interest in the ORC model, the CAMPD problem may be formulated. The resulting design problem is a mixed integer nonlinear problem (MINLP) which is prone to non-convergence for a number of reasons.

The ORC process model (equality constraints of the design problem) consists of a number of non-convex coupled equations, while property prediction models tend to be highly nonlinear. Further, as the choice of working fluid changes, across optimizations, many molecules may be infeasible with respect to the process model, which could result in non-convergence of optimizations (especially if simulator-based optimization is used). For example, a chosen molecule may have a critical point that is lower than the range of allowable pressures and temperatures and can thus not be condensed into a liquid in the cycle. In other scenarios, a molecule with a boiling point which is higher than that of the hot source cannot be obtained as a vapour at the outlet of the evaporator. An attempt to evaluate the vapour properties of such a stream can result in numerical failure.

To address the challenge of non-convergence and improve computational performance, we extend an algorithm [2, 3] which was recently developed for CAMPD of separation systems. The algorithm builds on the concept of embedding screening of molecules within an iterative MINLP optimizer [4, 5], such as Outer-Approximation [6]. Within this algorithm, each molecule generated by the master problem passes through a sequence of tests, before the evaluation of the primal problem for that molecule. The aim of the tests is to identify *a priori* regions within the process domain that are infeasible for a given molecule. If the entire domain is infeasible for a molecule, the molecule may be eliminated without evaluation of the primal problem, and information on the violated constraints are added to the master problem.Â For a molecule that passes the tests, the tests generate bounds on the feasible process domain for that molecule.

In this work, we develop tests specific to the design of ORC systems. The tests produce bounds on allowable operating pressures of the ORC for every feasible molecule. The tests also eliminate molecules whose phase behaviour (critical point, vapour pressure and so on) make them infeasible in the ORC. The algorithm is applied to solve three case studies of integrated ORC working fluid and process design. The SAFT-Î³ Mie [7] equation of state is used for prediction of thermodynamic properties of the working fluid. We find that the bounds obtained by the tests are useful in reducing the process domain and enable convergence of the primal problem. Without these bounds, the evaluation of the primal problem fails for certain molecules. The use of the algorithm enhances the robustness of the design problem, allowing one to obtain (locally) optimal solutions from a variety of starting points, which is useful as the problem is non-convex. Thus, the algorithm facilitates the use of rigorous process and property prediction models for the design of ORCs.

1 Adjiman, C. S., Galindo, A., and Jackson, G. (2014). Proceedings of the 8th International Conference on Foundations of Computer-Aided Process Design FOCAPD 2014. Elsevier

2 Gopinath S, Galindo A, Jackson G, Adjiman CS (2016). Proceedings of the 26th European Symposium on Computer Aided Process Engineering, edited by Kravanja Z, Computer Aided Chemical Engineering. Elsevier B.V.

3 Gopinath S, Jackson G, Galindo A, Adjiman CS (2016). AIChE Journal, 62:3484â€“3504

4 Buxton, A., Livingston, A. G., and Pistikopoulos, E. N. (1999). AIChE Journal, 45(4):817-843

5 Giovanoglou, A., Barlatier, J., Adjiman, C. S., Pistikopoulos, E. N., and Cordiner, J. L. (2003). AIChE Journal, 49(12):3095-3109

6 Duran, M. A. and Grossmann, I. E. (1986). Mathematical Programming, 36(3):307-339

7 Papaioannou, V., Lafitte, T., Aveñdano, C., Adjiman, C. S., Jackson, G., Müller, E. A., and Galindo, A. (2014). The Journal of Chemical Physics, 140(5)

497053 Integrated Computational Design of Cosmetic Products

Abstract:

Increased competition and globalized market dynamics have driven the accelerated growth of chemical industry [1]; consequently, different investigations have been directed in order to propose new resources in manufacture of high added-value products [2]. This work presents the implementation, in PYOMO®, of computational tool for integrated design of structured cosmetic products and its application in moisturizing lotion formulation.

The consumer-oriented products design requires a multiscale approximation, since the properties are closely related to molecular interactions that describe its microscopic and macroscopic structure [3]. Owing to the essential of multiscale observation approach, an integration of systemic procedures was performed in order to consider involved processes at different levels in product design; therefore, the tool was developed from an approximation defined by three modules:

- Surfactant Module: In charge of defining the surfactant (or mixture of surfactants) in order to specify molecular interactions that allows obtaining an appropriate surface tension and emulsion stability [4].
- Rheology Module: Defines the microstructure and macrostructure of emulsion from molecular scale with view to know its rheological properties [5].
- Consumer Module: Integrates a consumption preferences model based on sensory perception of customer and an economic model that aims at economic utility using market parameters [6].

In regard with previous ideas, this tool facilitates the design by projecting an appropriate formulation of cosmetic product, making it possible through an integrated optimization model raised to specify the variables of proposed modules, in quest of an optimal formulation, considering two objective functions according to user's requirements:

- Economic Function: Allows finding the product formulation and mass fractions per component focused on obtaining the maximum economic yield, taking into account the costs of manufacture, price and demand of the product.
- Preference function: Indicates the product formulation and mass fractions per component aimed at obtaining the maximum acceptance by consumer, according to established preferences parameters.

In conclusion, the implementation of this computational tool for integrated design of products is of great interest in the cosmetic industry; its importance lies in the optimization model that considers the interactions of proposed modules, for the purpose of generate a solid base in structured products formulation by simulations and thus reduce the resource spending in experimental design phases. Being a tool based on open source software with a flexible methodology brings with it certain advantages, it is not only possible to include new modules of interest according to the requirements of product design, but it also sets the basis for parallel developments of similar tools oriented to different Industries such as pharmaceuticals and food.

References

[1] B. V. Smith y M. G. Lerapepritou, «Integrative chemical product design strategies: Reflecting industry trends and challenges,» *Computers & Chemical Engineering,* vol. 34, nº 6, pp. 857-865, 2010.

[2] E. Cussler y G. Moggridge, Chemical Product Design, Cambridge: Cambridge University Press, 2011.

[3] R. Gani, «Chemical product design: challenges and opportunities,» *Computers & Chemical Engineering,* vol. 28, nº 12, p. 2441–2457, 2004.

[4] C. Wibowo y K. Ng, «Product-oriented process synthesis and development: Creams and pastes,» *AIChE Journal,* vol. 47, nº 12, pp. 2746 - 2767, 2001.

[5] M. Mattei, G. Kontogeorgis y R. Gani, «A comprehensive framework for surfactant selection and design for emulsion based chemical product design,» *Fluid Phase Equilibria,* vol. 362, nº 25, pp. 288 - 299, 2014.

[6] M. Bagajewicz, S. Hill, A. Robben, L. Heyde, M. Sanders, E. Sposato, C. Baade, S. Manora y J. Coradin, « Product Design in Price-Competitive Markets: A case study of a Skin Moisturizing Lotion,» *AIChE Journal,* vol. 57, nº 1, pp. 160 - 177, 2011.

497248 Modeling and Multi-Objective Optimization of an Industrial Ammonia Synthesis Process

Abstract:

Ammonia synthesis is a major important process in chemical industry. Ammonia application domain is very wide with major applications in manufacturing of agricultural fertilizers. Hence, it is particularly important for Canada where agricultural sector constitutes a significant part of national GDP. To keep up with continuously increasing demand, it is highly important to improve performance of existing facilities through process modeling, simulation and optimization. One of the way to achieve the goal is through mathematical modeling and optimization of operation and/or design parameters of a plant. Due to large scale of production, even minor improvements can result in significant profits or savings.

The conventional synthesis is based on Haber process - reaction of gaseous nitrogen and hydrogen over a solid catalyst to produce ammonia. The reaction is carried out at elevated temperature (400-500 °C) and pressure (80-230 atm.). The feedstock comes from a steam reforming unit at the same plant and is made of stoichiometric mixture of nitrogen and hydrogen with a minor portion of inert. The synthesis reaction occurs in one or few (usually, 1-4 beds) fixed catalytic beds. Since the process is exothermic, the vast amount of heat is generated and re-used in the process. In this work, a collaborative research was conducted with a large ammonia producing plant in Canada aiming to investigate possibilities of further improving ammonia production and heat duty in the ammonia converter. First-principle mathematical modeling of gas-solid catalytic reaction and heat transfer equipment was performed. A simple enough yet accurate model was developed to adequately simulate the ammonia converter. Main features of the model are one dimensional plug flow with Fick-type diffusion of reactant inside solid catalyst particle.

Importantly, kinetic parameters of catalyst are being the part of proprietary information, and hence, were not available. The model was first tuned with industrial data to provide estimate for catalyst kinetic parameters. For the proper estimation of kinetic parameters, a large set of industrial data was used covering time span over 12 years. Since data set is large and is full of information irrelevant to normal steady state operation, which included not only scheduled maintenance, but also short term disruptions, cold and hot start up and shutdown, sensor failures, etc., and hence, it was not evidently clear which data to use for the model validation. Hence, a data pre-treatment procedure based on data clustering around steady-states was performed to filter out relevant industrial data points. Thereafter, the model was fitted over several different time-dependent groups of data. It was discovered that even though catalyst life time for ammonia catalyst is large (> 10 years), the catalyst still experiences deactivation and it should be accounted at different stages of operation. Incorporating all the above, the model was validated over industrial data and showed good consistency with observed values. Finally, the validated model was used to perform multi-objective optimization with modified Non-Dominated Sorting Genetic Algorithm (NSGA-II). The primary goals were to maximize production of ammonia, minimize heat requirement for the reaction and minimize total catalyst loading in converter. Three optimization case studies were performed, where objectives were treated pair-wise to provide a Pareto optimal solution for converter operation.

In the first case, optimal values for three process parameters - namely, converter pressure, temperature of feed stream and ratio of quench stream to significantly reduce heat duty while keeping ammonia production above industrially acceptable level. In the second case, a different catalyst distribution among beds, was proposed to allow for higher ammonia production. Three cases with different distribution were obtained where total catalyst loading in the converter is lower, equal and higher than current. All three cases showed higher ammonia production compared to the base case. Additionally, a change in optimal conditions with respect to time-on-stream for the catalyst has been addressed and guidelines for the converter operation were provided like the first case.

497475 Generation of Sustainable Hybrid Process Flowsheets Using Procafd

Abstract:

The chemical industry is always looking for innovative process designs, which are both efficient and sustainable. To generate an optimal flowsheet, what is needed is a systematic way to identify the types of tasks-operations that need to be performed, the corresponding design of the operation-equipment, their configuration, mass-energy flows etc. Due to the fact that process synthesis problems are by nature combinatorial and open ended, a number of different solution approaches have been proposed. However, the solution for any synthesis-design problem is dependent on the search space of alternatives and the process performance criteria which in most cases are influenced by economic factors. This work focuses on the development of a computer aided tool (ProCAFD), capable of enumerating the entire feasible search space, analyze and determine the most sustainable process. ProCAFD is based on decomposition based method developed by Tula et al. (2015). On a higher level the workflow of ProCAFD can be divided into three stages: 1) Synthesis stage â€" Here a hybrid approach based on group contribution is used to generate and rank the alternatives. In this stage, chemical process flowsheets are synthesized in the same way as atoms or groups of atoms are synthesized to form molecules in computer aided molecular design (CAMD) techniques. The building blocks in flowsheet synthesis are called process-groups, which represent a single or set of unit operations that are selected by employing a thermodynamic insights based method. These building blocks are then combined using connectivity rules to generate all the feasible flowsheet alternatives. The main advantage of representing the flowsheet with process-groups is that, the performance of the entire process can be evaluated from the contributions of the individual process-groups towards the selected flowsheet property (for example, energy consumed). Available flowsheet property models include energy consumption, carbon footprint, product recovery, product purity etc. In this way, the entire list of feasible chemical process flowsheets is quickly generated, screened and selected. 2) Analysis stage â€" In this stage, comprehensive investigation consisting of sustainability, life cycle and economic analysis is performed to identify areas of improvement for the generated process alternative. The identified areas or process hotspots are translated into design targets that, if satisfied will guarantee process improvement. 3) Innovation stage â€" In this stage different innovative design strategies are used to match the set design targets. Different design strategies include simultaneous process optimization and heat integration where determination of optimal operating conditions and design parameters is solved together with heat integration involving minimizing utility consumption. One other strategy ProCAFD uses is a generalized method based on Tula et al. (2016) to replace distillation columns in the selected design with a hybrid distillation-membrane system to reduce the operational cost substantially. Other strategies include process intensification, where the phenomena based intensification method developed by Babi et al. (2015) is employed to generate intensified processes.

In this work, the software implementation of the developed methodology will be highlighted through several examples involving chemical and biochemical processes. In all the cases studied, the developed software could generate innovative and more sustainable solutions very quickly.

References:

A. K. Tula, B. Befort, N. Garg, K. V. Camarda, R. Gani, Sustainable Process Design & Analysis of Hybrid Separations, Computers & Chemical Engineering (2016). http://dx.doi.org/10.1016/j.compchemeng.2016.11.031.

A. K. Tula, M. R. Eden, R. Gani, 2015, Process synthesis, design and analysis using a process-group contribution method, Computers & Chemical Engineering, 81, 245-259.

D. K. Babi, J. Holtbruegge, P. Lutze, A. Gorak, J.M. Woodley, R. Gani, 2015, Sustainable Process Synthesis-Intensification, Computers & Chemical Engineering, 81, 218-44.

499195 Total Site Integration As a Synthesis Tool to Select Biomass Valorization Paths and Schedule Multiple-Feedstock Operations

Abstract:

This paper presents a systematic approach to select and integrate product portfolios in multiple-feedstock biorefineries. Value chains include all potential routes from biomass to intermediate and end-products that may enter biorefineries. Once the problem requires minimum energy requirements to secure process sustainability, Total Site integration is called to minimize consumption of steam (heat transfer mean among processes). The use of Total Site Analysis to screen biorefinery paths has been recently presented by the authors (Kokossis et al., 2015) to detect promising process portfolios that optimally integrate each other in terms of energy use. The methodology introduced a modified transshipment model to evaluate energy synergies among candidate processes of value chains given the type and annual capacity of biomass feedstock. Once the majority of applications refer to lignocellulosic biorefineries, the analysis naturally extends to the assessment of changes in operation and efficiencies due to seasonal availability of each lignocellulosic biomass type. This paper extends previous concepts to assess the impacts of switching feedstocks on energy efficiencies and the selection of biorefinery portfolios.

Lignocellulosic biorefineries are based on the exploitation of the three fundamental chemical components: C5 and C6 sugars and lignin. Though products of value chains remain the same with changes in the type of lignocellulosic material – wheat straw, rice, poplar or birch, they all include sugars and lignin – changes appear in process capacities and energy consumption due to seasonal variations in capacities and yields in sugars and lignin for each biomass type. Accordingly, heat interactions and recovery among processes based on the valorization of each component change by switching feedstocks during the year.

The combinatorial process synthesis and integration problem is formulated by the mass balances along value chains (selection of paths) and energy balances at each interval of the proposed transshipment (evaluation of selected paths). The problem also assumes changes in biomass capacities at discrete time periods for each candidate type of biomass. Accordingly, the domain of flowrates along value chains is expanded by the indexes of biomass type and the time period that is applied. Large deviations in common chemicals flows and processes capacities at different periods are not industrially accepted. Likewise, the domain of heat flows in the transshipment model and utilities is also expanded by time and biomass index and limited for logic deviations. The MILP model combines selection of value chain paths with integration in transshipment model to evaluate energy savings at each period and thus, to synthesize the seasonal feedstocks plan and select the processes portfolio that minimize energy cost of the biorefinery site.

References

A.C. Kokossis, M. Tsakalova, K. Pyrgakis, Design of integrated biorefineries, Computers & Chemical Engineering, 2015, 81, 40–56

AUTHOR INDEX

AUTHOR INDEX

AUTHOR INDEX

AUTHOR INDEX

AUTHOR INDEX

AIChE
120 Wall Street, FL 23
New York, NY 10005-4020

ISBN 978-1-5108-5800-8

9 781510 858008

CURRAN ASSOCIATES INC.
proceedings.com